Springer Theses

Recognizing Outstanding Ph.D. Research

Aims and Scope

The series "Springer Theses" brings together a selection of the very best Ph.D. theses from around the world and across the physical sciences. Nominated and endorsed by two recognized specialists, each published volume has been selected for its scientific excellence and the high impact of its contents for the pertinent field of research. For greater accessibility to non-specialists, the published versions include an extended introduction, as well as a foreword by the student's supervisor explaining the special relevance of the work for the field. As a whole, the series will provide a valuable resource both for newcomers to the research fields described, and for other scientists seeking detailed background information on special questions. Finally, it provides an accredited documentation of the valuable contributions made by today's younger generation of scientists.

Theses may be nominated for publication in this series by heads of department at internationally leading universities or institutes and should fulfill all of the following criteria

- They must be written in good English.
- The topic should fall within the confines of Chemistry, Physics, Earth Sciences, Engineering and related interdisciplinary fields such as Materials, Nanoscience, Chemical Engineering, Complex Systems and Biophysics.
- The work reported in the thesis must represent a significant scientific advance.
- If the thesis includes previously published material, permission to reproduce this must be gained from the respective copyright holder (a maximum 30% of the thesis should be a verbatim reproduction from the author's previous publications).
- They must have been examined and passed during the 12 months prior to nomination.
- Each thesis should include a foreword by the supervisor outlining the significance of its content.
- The theses should have a clearly defined structure including an introduction accessible to new PhD students and scientists not expert in the relevant field.

Indexed by zbMATH.

More information about this series at http://www.springer.com/series/8790

Thomas Day Goodacre

Applied Laser Spectroscopy for Nuclear Physics

Isotope Shifts in the Mercury Isotopic Chain and Laser Ion Source Development

Doctoral Thesis accepted by
University of Manchester, Manchester, UK

Springer

Author
Dr. Thomas Day Goodacre
TRIUMF
Vancouver, BC, Canada

Supervisor
Prof. Jon Billowes
School of Physics and Astronomy
University of Manchester
Manchester, UK

ISSN 2190-5053 ISSN 2190-5061 (electronic)
Springer Theses
ISBN 978-3-030-73888-4 ISBN 978-3-030-73889-1 (eBook)
https://doi.org/10.1007/978-3-030-73889-1

This Springer imprint is published by the registered company Springer Nature Switzerland AG
The registered company address is: Gewerbestrasse 11, 6330 Cham, Switzerland

Supervisor's Foreword

The Ph.D. thesis of Thomas Day Goodacre describes his work on developments of the Resonance Ionization Laser Ion Source (RILIS) at the ISOLDE radioactive ion beam facility at CERN. This is one of the most advanced and versatile ion sources of its type in the world and supports research programmes of many of the international community working at ISOLDE. As the research progressed he began to focus on improvements specifically for the mercury isotopes leading to the crowning achievement of the thesis of fully mapping the nuclear shape evolutions for the exotic neutron-deficient mercury isotopes.

Laser ion sources such as RILIS use step-wise laser-excitations of an atom to reach photo-ionization. When one or more of these steps is resonant with a tuned monochromatic laser, it is possible to obtain a high degree of element selectivity and still maintain a good efficiency of the ion source. When this is coupled with isobar selection by a magnetic separator, beams of a single nuclide can be obtained with low isobaric contamination. If the resonant step laser is frequency-tuned while monitoring the separated beam flux, then very sensitive measurements of isotope shifts can be carried out "in-source".

There were two key stages in the development work for mercury. Tom's primary task of developing ionization schemes to expand the range of elements available from RILIS provided new schemes for five elements: chromium, tellurium, germanium, mercury and radium. Of special significance was the 60-fold efficiency improvement for mercury.

The second stage was to make a laser ion source that was compatible with the molten lead target used for the production of mercury radioisotopes. The solution, called VADLIS (Versatile Arc Discharge and Laser Ion Source), used the molten lead target connected to a FEBIAD-type arc discharge source which acted as the laser-atom interacting region. Tom was instrumental in the development of this new ion source and performed extensive off-line and on-line tests for several elements. Following the completion of these two stages, it was then possible to apply the in-source spectroscopy technique to extend the mercury isotope shift measurements.

The long-standing interest in the light mercury isotopes comes from the huge odd-even staggering of the nuclear mean square charge radii seen four decades ago in optical isotope shift measurements of the neutron-deficient isotopes down to ^{181}Hg. This is attributed to the large prolate deformations of the odd-neutron isotopes ^{181,183,185}Hg but it was not known how far below the neutron mid-shell (at ^{184}Hg) this dramatic shape -coexistence would persist. The new measurements were carried out by a collaboration of 16 institutes co-led by Tom. They were magnificently successful and extended to study down to ^{177}Hg (127 ms half-life). They provided the definitive answer that the shape staggering stops at ^{180}Hg, below which all isotopes have returned to the near-spherical shape observed for isotopes above ^{186}Hg.

Overall, the legacy of Tom's research has been to provide an ion source with enhanced performance and a range of elements which are of immediate benefit to the ISOLDE physics community and hold potential benefits for the next generation of higher intensity radioactive beam facilities.

Manchester, UK
September 2018

Prof. Jon Billowes

Abstract

This work centres around the development and application of the Resonance Ionization Laser Ion Source (RILIS) of the ISOLDE radioactive ion beam facility based at CERN. It is based on a Ph.D thesis from 2017, the status of the systems discussed are correct as of that date. The RILIS applies step-wise resonance photo-ionization, to achieve an unparalleled degree of element selectivity, without compromising on ion source efficiency. Because of this, it has become the most commonly used ion source at ISOLDE, operating for up to 75% of ISOLDE experiments. In addition to its normal application as an ion source, the RILIS can be exploited as a spectroscopic tool for the study of nuclear ground-state and isomer properties, by resolving the influence of nuclear parameters on the atomic energy levels of the ionization scheme.

There are two avenues of development by which to widen the applicability of the RILIS: laser ionization scheme development, enabling new or more efficient laser ionized ion beams and the development of new laser-atom interaction regions. New ionization schemes for chromium, tellurium, germanium, mercury and radium have been determined. Additionally, for the first time, the anode cavity of the Versatile Arc Discharge Ion Source (VADIS), ISOLDE's variant of the Forced Electron Beam Induced Arc Discharge Ion Source (FEBIAD) type ion source was used as the laser-atom interaction region. A new element selective RILIS mode of operation was established, enabling the ISOLDE RILIS to be coupled with molten targets for the first time, increasing the flexibility of ISOLDE operation and opening a direction for future developments. This combined ion source was termed the VADLIS or Versatile Arc Discharge and Laser Ion Source.

A combination of the developments presented in this thesis: an improvement of the RILIS ionization efficiency of mercury and the ability to couple the RILIS with molten targets, satisfied the pre-requisites for the long-awaited extension of the laser spectroscopy studies of exotic mercury isotopes. A sudden onset of extreme shape staggering in the neutron deficient mercury isotopes was revealed by optical pumping and laser spectroscopy experiments at ISOLDE in the 1970s and 1980s, with measurements conducted down to ^{181}Hg. Despite this being one of the most remarkable examples of shape coexistence in the nuclear chart, in the intervening decades, the cessation point of this odd-even staggering had yet to be unambiguously determined through measurements of nuclear ground state charge radii. This open question was successfully resolved using the ISOLDE RILIS for in-source resonance ionization spectroscopy. The experiment was performed as part of a large collaboration, using the KU Leuven Windmill system for α-detection; direct ion counting with the ISOLTRAP multi-reflection time of-ight mass spectrometer (MR-ToF MS); and ion beam current measurements using the ISOLDE Faraday cups. The sensitivity of the technique enabled the measurements to be extended down to ^{177}Hg, providing a definitive answer, that the extreme shape staggering stops at ^{180}Hg. In addition to extending the measurements at the neutron deficient end of the mercury isotope chain, the relative mean square charge radii of both ^{207}Hg and ^{208}Hg were determined. This extended the measurements beyond the N = 126 shell closure, enabling the characterization of the "kink" in the trend of the isotope shifts.

Acknowledgements

There are a huge number of people who have been instrumental to the work presented in this thesis.

Starting closest to the work, I first have to thank Bruce for being a constant source of encouragement, support and enthusiasm as a supervisor. The ISOLDE RILIS team over the last 4 years (Valentin, Bruce, Sebastian, Ralf, Daniel, Christoph, Kati, Dima, Anatoly, Maxim, Pavel, Nobu, Julia, Annie, Sasha and Pierre) has been a fantastic place to work and learn, it's been a pleasure to get to know all you. Despite demanding requirements for on-line operation it feels like we have been able to achieve a lot. Additionally, I have to thank Anatoly for all of his help and assistance with the data analysis. To my supervisor Jon and co-supervisors Kieran and Thomas at Manchester, thank you for all of the advice, support and wine over the last few years. To my examiners, Iain and Paul, thank you for the feedback on this work. Finally, a big thank you to the whole in-source RIS collaboration, they've been great experiments to be involved in.

Closer to home, beginning in Geneva (now Vancouver) a massive thank you to Maria for putting up with the nights and weekends I've spent at CERN or on call, together with weeks of travelling. Additionally for the amount of time (and money) spent joining me in Geneva coffee shops during the "free" weekends! Moving further from CERN and Geneva, a huge thanks to my Mum for the unwavering support over the years. I've always had a home to go back to in Manchester, which has made life much easier when moving away as it's felt like I've never left. To my Nan, Uncle Howard and Auntie Sarah, thank you all for your help, understanding and short notice availability during my far too brief day trips to Oxford from Geneva. To all of the friends I've barely seen over the last couple of years, I look forward to seeing you soon!

Finally, thank you to Carsten, Rob and Magda. I've not forgotten the requirements of LA^{3}NET "this project has received funding through the European Union's Seventh Framework Programme for Research, Technological Development and demonstration under Grant Agreement 289191". It was a lot more than just funding though, the people involved in LA^{3}NET and the opportunities provided by the Marie Curie Fellowship, made it an incredible 3 years.

Contents

Chapter 1
Introduction

Abstract This chapter is an introduction to the thesis. The basics of radioactive ion beam production and resonance ionization spectroscopy are discussed. Following this, the motivation is summarised for the ion source development in Chaps. 6 and 7 and the study of exotic mercury isotopes in Chaps. 8–10.

1.1 Fundamentals

Nuclear physics plays a key part in modern life. From ionization smoke alarms, electricity production or the optimization of manufacturing processes to medical imaging and therapy, carbon dating or nuclear forensics, many of the mature technologies that are used daily are spin-offs from fundamental research and the technology and techniques developed for it. Detailed discussions and numerous examples can be found in [1, 2]. To study nuclear properties or phenomena, reactions typically have to be induced or radioactive isotopes have to be produced. Today, one of the principal options for studying radioactive isotopes is at radioactive ion beam (RIB) facilities.

RIB facilities have been described by colleagues in various ways: from "a playground for alchemists" to "exotic isotope factories". Fundamentally, they exploit nuclear reactions to create rare and short-lived isotopes of elements: nuclei where the ratio of protons and neutrons results in an instability of the nucleus. As an example, the combination of 79 protons (Z) and 118 neutrons (N) corresponds to a nucleus of $^{N+Z=197}$Au the stable isotope of gold. Adding or removing a neutron, results in $^{N+Z=198}$Au or $^{N+Z=196}$Au, still isotopes of gold but with half-lives of 2.6941 (2) days and 6.1669 (6) days respectively [3]. Removing 26 neutrons results in a nucleus of $^{N+Z=171}$Au, a gold isotope with a half life of just 22.6 (24) μs [3]. RIB facilities provide scientists with the opportunity to conduct experiments on or with these exotic isotopes. These experiments can range from the investigation of promising new cancer treatments [4] to nuclear astrophysics [5].

Lasers play an increasingly integral role at such facilities, both in the production of radioactive ion beams and the study of the exotic isotopes themselves. Considering

T. Day Goodacre, *Applied Laser Spectroscopy for Nuclear Physics*, Springer Theses,
https://doi.org/10.1007/978-3-030-73889-1_1

light as packets of energy (photons), the colour of the light corresponds to the energy contained within the photon. One of the advantageous properties of lasers is their ability to produce monochromatic light or, equivalently, photons with a very narrow range of energies. Spectral absorption lines correspond to the excitation of an atomic electron and the excitation energies are unique to a particular element. For a given transition, a photon energy matching the excitation energy is required in order to induce the excitation. Thus, photons with a matching energy could be considered like a key to a lock, with a specific photon energy required to excite an atomic electron of a certain element. Lasers can therefore be used to target a specific atomic absorption line, promoting an electron to an excited energy level.

A commonly used laser-atom interaction region at RIB facilities is a metal tube with an internal diameter of 3 mm. Taking typical parameters, laser light with an average power of 3 W, a wavelength of 580 nm and a spectral line width of 0.01 nm is directed into such a metal tube to target an atomic transition. The specific requirement of a laser for this role can be better understood by a comparison to sunlight. The spectral irradiance of the sun at sea level over this wavelength range is $\approx 1.1\ \mathrm{Wm^{-2}nm^{-1}}$ [6], or 39,000,000 times lower than the irradiance of the laser over the 0.01 nm spectral range, directed into the metal tube.

It is possible to use multiple lasers to target a sequence of atomic lines, here the analogy shifts from that of a lock and a key, to a safe where a unique combination is required to unlock it. With a sufficient number of sequential excitation steps, it is possible transfer enough energy to an atomic electron for it to break free of the Coulomb attraction of the nucleus, liberating it and transforming the neutral atom into a charged particle (a singly charged ion). This method of ionization enables an element of interest to be selectively ionized, by applying the combination of excitation steps that are unique to it. The ions can then be extracted from the cloud of reaction products produced in the nuclear reactions at the RIB facilities. The ionized particles can be manipulated with electric and magnetic fields, to enable their transport to experimental set-ups and the experiments to take place.

Atomic transitions are, to a varying degree, sensitive to nuclear properties. Therefore the atomic excitations can also be used as probes to study the nucleus. By varying the energy of the photons used to target a particular atomic transition, it is possible to determine part-per-million shifts in the electronic energy levels, enabling attometer-scale fluctuations in the nuclear charge radius to be identified. Such laser spectroscopy techniques enable the evolution of ground state nuclear structure to be charted across the nuclear landscape.

1.2 Radioactive Ion Beam (RIB) Facilities

If placed over a Venn diagram of scientific disciplines, RIB facilities would lie rather close to the centre. The production of RIBs first requires plasma and accelerator physics to produce the "driver beam". The driver beam is then impacted upon a target, where nuclear physics, materials science and chemistry are required to optimize the production and extraction of the isotopes of interest. Following extraction from the target, atomic, laser and plasma physics, coupled with chemistry and materials science may be employed if an ion source is used. Finally, accelerator physics is once again required as the RIB is extracted and delivered to an experiment. The experiments themselves then range from nuclear astrophysics or probes of the standard model to chemistry, nuclear medicine, biology and materials science [7].

The first experimental setup for producing RIBs was reported by Kofoed-Hansen and Nielson in 1951 [25]. They had created the first ISOL (Isotope Separator On-Line) type facility, coining the ISOL acronym in their publications describing its use, as the coupled and simultaneous operation of the accelerator delivering the incident projectiles for radionuclide production and an isotope separator. The most up to date cross-section of RIB technology can be found in the proceedings of the triennial EMIS (international conference on ElectroMagnetic Isotope Separators and related topics) conference [8]. Similarly, a representative cross-section of the physics undertaken at RIB facilities can be found in the proceedings of the triennial ARIS (Advances in Radioactive Isotope Science) conference [9]. Additionally, there are numerous review articles and published lecture notes on the topic ([7, 10]).

As a first approximation, the yield of a RIB of an isotope of interest can be expressed as

$$Y_{(^N X)} \propto \sigma \Phi n v \eta, \tag{1.1}$$

where $Y_{(^N X)}$ is the yield of the isotope of interest (s^{-1}), σ is the reaction cross-section (cm^2) which is driver beam energy dependent, Φ is the the driver beam particle flux ($cm^{-2}s^{-1}$), n is the number of target atoms per unit volume (cm^{-3}), v is the interaction volume (cm^3) and η is the efficiency of extraction as an ion beam. A number of the factors are interdependent: the energy of the driver beam degrades and diverges along the length of a target which respectively changes the reaction cross-section and reduces the flux. Increases in the interaction volume may increase the production yields, however, it may also increase the release time for reaction products which would then decrease the yields of comparatively short-lived isotopes. η is a combination of multiple efficiencies that differ between types of RIB facilities. These can include: the target release efficiency, the transport efficiency between the target and the ion source, the ionization efficiency, the ion extraction efficiency and the RIB transport efficiency.

This work will focus on studies conducted at the ISOLDE facility [11] based at CERN, the "low energy" (30-60 keV) beamlines of the ISOLDE facility are shown in Fig. 1.1.

ISOLDE is a thick target type ISOL type facility, which means that the target is of sufficient thickness to stop and thermalize the reaction products resulting from the impinging driver beam. Other types of RIB facilities include In-Flight facilities [13, 14] where the reaction products are directly ejected from a "thin" target, retaining much of the driver beam energy and hybrids which use gas catchers combined with thin targets [15–17]. At thick target facilities, the half-life of the isotope of interest is particularly relevant to the release efficiency, as the release time can stretch to

Fig. 1.1 The low energy beamlines of the ISOLDE facility based at CERN. Image reproduced from the CERN document server [12]

minutes or even hours, compared to the time for ionization, ion extraction and ion beam transport which combined can be of the order of a few hundred milliseconds.

1.3 Nuclear Physics at Thick Target ISOL Facilities

There are a number of comprehensive overviews of the nuclear physics and other research conducted at thick target ISOL facilities including books and published lecture series [18–22]. A recent example of the science undertaken can be found in the 2017 *Focus on Exotic Beams at ISOLDE: A Laboratory Portrait* published in Journal of Physics G: Nuclear and Particle Physics [23]. Laser spectroscopy experiments study nuclear ground state and isomer properties via the influence of the nucleus on atomic electrons. Mass measurements investigate the binding energy of exotic nuclei. Decay spectroscopy experiments enable investigations ranging from the determination of decay modes of exotic nuclei, to weak interaction studies or fast timing spectroscopy. Coulomb excitation and transfer reaction experiments investigate nuclear properties from excited states to reaction channels and cross-sections.

Fig. 1.2 The ISOLDE RILIS laser ion source. The system is described in detail in Sect. 5.2.2. Image reproduced from the CERN document server [12]

1.4 Resonance Ionization Spectroscopy (RIS) at Thick Target ISOL Facilities

Resonance ionization spectroscopy (RIS), discussed in Sect. 1.1, involves the step-wise resonance excitation and ionization of atoms of the element of interest. RIS is typically applied for two purposes at thick target ISOL facilities. The most common application is for ion beam production, applying a resonance ionization laser ion source (RILIS) to ionize atomic reaction products following effusion from the target material. The combination of a RILIS with mass separation enables isotope specific purification. The RILIS of the ISOLDE facility [24] is typically used for 70–80 % of ISOLDE experiments. The principles of RILIS are explored in Chap. 4 and the ISOLDE RILIS is described in Sect. 5.2.2. The ISOLDE RILIS is shown in Fig. 1.2.

The second application is for nuclear structure studies, exploiting the interaction of the nucleus with the atomic electrons. Through these investigations, the nuclear spin, charge radii, dipole and quadrupole moments of isotopes can be investigated, the underlying theory is discussed in Chap. 3. RILIS can be applied for these studies, termed "in-source spectroscopy experiments". These experiments are limited to the heavier region of the nuclear chart, the reasons for this are discussed in Sect. 4.2.2. Alternatively RIS can be applied collinearly on an accelerated ion or atom (following neutralization) beams. This technique is regularly applied at ISOLDE on the collinear resonance ionization spectroscopy (CRIS) beam line [25]. The techniques are briefly compared in Sect. 4.2.2.

1.5 This Work: Ion Source Development and Laser Spectroscopy of the ISOLDE RILIS

Two of the directions by which to expand the applicability of the ISOLDE RILIS are: laser ionization scheme development enabling new or more efficient laser ionized ion beams, or the development of new laser-atom interaction regions. The results of investigations into new RILIS ionization schemes for chromium (Sect. 7.2), germanium (Sect. 7.3), mercury (Sect. 7.4) radium (Sect. 7.5) and tellurium (Sect. 7.6) are presented. In addition, a new laser atom interaction region has been investigated, coupling the RILIS with the VADIS, ISOLDE's variant of the FEBIAD type arc discharge ion source (Chap. 6). A new element selective RILIS mode of operation has been established, through off-line testing with gallium and barium and on-line experiments with mercury, cadmium and magnesium. This development enabled the ISOLDE RILIS to be coupled with molten targets for the first time, increased the flexibility of ISOLDE operation and opened a direction for future developments.

1.6 This Work: Mercury Isotopes "the Playground of the Optical Spectroscopist"

Mercury isotopes were described as the playground of the optical spectroscopist by G. Fricke and K. Heilig in their compendium of nuclear charge radii data [26], an assessment based on the extensive history of optical studies of these isotopes. This Section presents a brief overview of the previous hyperfine structure (HFS) investigations of mercury isotopes and the motivation for further studies. Much of the history is summarized in [27, 28].

1.6.1 Measurements Pre-1972

The targeting of the 254 nm mercury line for isotope specific ionization was first reported in the early 1930s by K. Zuber [29], applying techniques pioneered by S. Mrozowski [30]. The advent of synthetic isotope production and the first generations of RIB facilities increased the range of available isotopes [31, 32], while improved interferometric techniques enabled increasing precision in the measurement of the HFS and isotope shifts [33]. A compilation of the measurements can be found in [28]. The status of the mercury isotope shift measurements in 1971 is presented in Fig. 1.3.

1.6.2 Measurements 1972–1986

A sudden jump in the trend of charge radii between ^{187}Hg and ^{185}Hg was reported by Bonn et al. [34] using a pioneering pre laser β-RADOP (β-radiation detected optical pumping) technique, capable of studying isotopes with nuclear spin $I \neq 0$. The addition of their results is presented in Fig. 1.4a and the technique is outlined

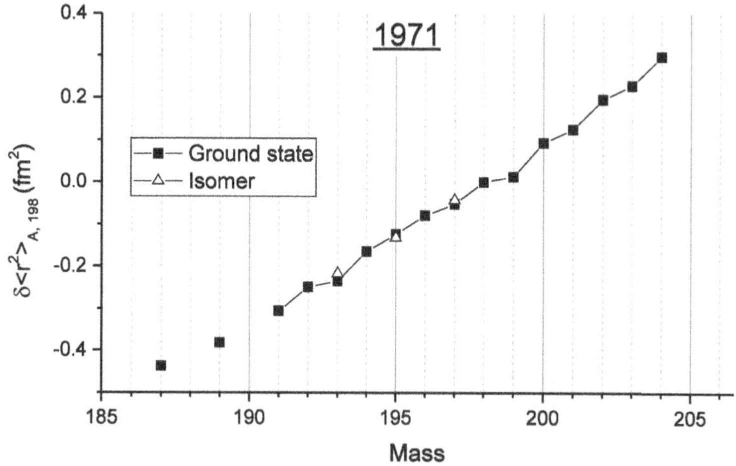

Fig. 1.3 Isotope shifts of mercury isotopes measured before 1972. Data taken from [28]

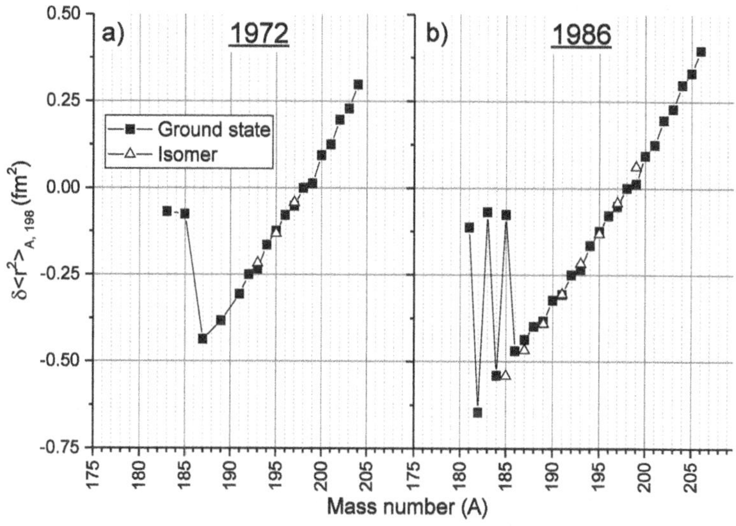

Fig. 1.4 The status of the isotope shift measurements of mercury isotopes in 1972 (**a**) and 1986 (**b**). Data taken from [28]

in [35]. Various explanations for the magnitude of the isotope shift were suggested, including bubble nuclei [36] and an increase in the proton surface density [37]. The phenomenon was correctly interpreted by [38–43] as a variation in the shape of the nuclei between a weak oblate deformation and a strong prolate deformation.

γ-spectroscopy identified an intruder rotational band, responsible for the shape coexistence and indicated that the ground state of ^{184}Hg corresponded to the band related to the weakly oblate deformation, therefore suggesting an odd-even shape staggering between ^{186}Hg and ^{184}Hg [44, 45]. This was confirmed optically by Kühl et al. [46], using laser spectroscopy which enabled an extension of the optical measurements to the $I = 0$ even isotopes.

The first "satisfactory" [27] explanation of the shape staggering was provided by Frauendorf and Pashkkevich [43] who modelled the nuclei by applying an approach combining the liquid drop model with Struinsky's shell correction [47]. They suggested that the shape staggering was the result of a differing loss in pairing energy between the two competing minima, following the addition of an unpaired neutron blocking the last Fermi level. The discrepancy in the pairing energy between the oblate and prolate minima was explained as the result of the differing densities of the single particle neutron levels for either the weakly deformed oblate, or the strongly deformed prolate deformations. The two minima are understood to be close (within a few hundred keV of each other), consequently the blocking effects are sufficient to tip the balance between the minima for the odd-A nuclei, resulting in the shape staggering.

Laser spectroscopy measurements extended the chain of $\delta\langle r^2\rangle^{198,A}$ measurements to span between ^{182}Hg and ^{206}Hg by 1986 [28, 48, 49], including the identification of the ≈24 GHz isomer shift in ^{185}Hg [50]. The status of the mercury isotope shift measurements by 1986 is presented in 1.4b.

The sudden and pronounced isotope and isomer shifts in neutron deficient mercury are considered to have made a significant contribution to the discovery of shape coexistence, as they provided a direct demonstration of the phenomenon [51]. It was described by K. Heyde and J.L. Wood as "one of the most remarkable discoveries in nuclear structure physics in the last 50 years" [52].

1.6.3 Motivation for Extending the Optical Measurements

The optical $\delta\langle r^2\rangle^{198,A}$ measurements presented in Fig. 1.4b indicate two opportunities for further investigation:

- At the neutron deficient end of the mercury isotope chain, the cessation point of the odd-even shape staggering had yet to be definitively determined through the study of nuclear ground state properties. Further in-beam studies, revealed an increase in the level energies of the prolate intruder structure with decreasing N from ^{182}Hg [53]. The return to a weakly deformed shape for the odd mercury isotopes below A=181 was suggested based on γ-spectroscopy [54, 55] and α-spectroscopy results [56]. Theoretical modelling was in agreement with the level energies of the prolate intruder structure [57] and a general reduction in deformation [58]. However, there was no definitive determination of whether the shape staggering would continue beyond ^{180}Hg.

- At the neutron rich end of the isotope chain, the experimentally measured isotopes extended as far as the N=126 shell closure. An extension of these measurements would provide information on the nuclei following the shell closure and the characteristic "kink" in the charge radii.

1.6.4 In-Source Laser Spectroscopy of Mercury Isotopes

The results of a study to extend the optical measurements of the mercury isotopic chain further into the neutron rich and neutron deficient regions of the nuclear chart [59] are presented in this work. Chapter 8 describes the in-source spectroscopy experiment which provided answers to these questions. Chapters 9 and 10 discuss the results for the neutron deficient and neutron rich measurements, respectively, in the context of recent experimental work, theoretical predictions and the charge radii systematics of the region. These measurements extend the isotope shift measurements of the mercury chain to span from ^{177}Hg to ^{208}Hg, the longest unbroken chain of optically determined isotope shifts in the nuclear chart. The ISOLDE community at the ISOLDE Workshop four months prior to this experiment is presented in Fig. 1.5.

Fig. 1.5 The ISOLDE community four months prior to the in-source laser spectroscopy experiment. Image reproduced from the CERN document server [12]

References

1. Committee on the Assessment of and Outlook for Nuclear Physics (2013) Nuclear physics: exploring the heart of matter. National Academies Press, Washington DC. https://doi.org/10.17226/13438
2. Hayes AC (2017) Applications of nuclear physics. Rep Progr Phys 80(2):026301. https://doi.org/10.1088/1361-6633/80/2/026301
3. Audi G, Kondev FG, Wang M, Huang WJ, Naimi S (Mar 2017) The NUBASE2016 evaluation of nuclear properties. Chinese Phys C 41(3):030001. https://doi.org/10.1088/1674-1137/41/3/030001
4. Müller C, Vermeulen C, Köster U, Johnston K, Türler A, Schibli R, van der Meulen NP (2017) Alpha-PET with terbium-149: evidence and perspectives for radiotheragnostics. EJNMMI Radiopharm Chem 1(1):5. https://doi.org/10.1186/s41181-016-0008-2
5. Wolf RN, Beck D, Blaum K, Böhm C, Borgmann C, Breitenfeldt M, Chamel N, Goriely S, Herfurth F, Kowalska M, Kreim S, Lunney D, Manea V, Minaya Ramirez E, Naimi S, Neidherr D, Rosenbusch M, Schweikhard L, Stanja J, Wienholtz F, Zuber K (2013) Plumbing neutron stars to new depths with the binding energy of the exotic nuclide 82Zn. Phys Rev Lett 110(4):041101. http://link.aps.org/doi/10.1103/PhysRevLett.110.041101
6. Mecherikunnel AT, Gatlin JA, Richmond JC (May 1983) Data on total and spectral solar irradiance. Appl Opt 22(9):1354. https://doi.org/10.1364/AO.22.001354
7. Huyse M (2004) The why and how of radioactive-beam research, pp 1–32. http://link.springer.com/10.1007/978-3-540-44490-9_1
8. Bollen G, Mittig W, Morrissey D, Schwarz S, Villari A (2015) Proceedings of the XVIIth international conference on electromagnetic isotope separators and related topics (EMIS2015), Grand Rapids, MI, USA, 11–15 May 2015. Nuclear instruments and methods in physics research section B: beam interactions with materials and atoms 376:1–2, 2016. https://doi.org/10.1016/j.nimb.2016.04.053
9. Zenihiro J, Yako K, Doornenbal P, Tohru M (eds) (2015) Proceedings of the conference on advances in radioactive isotope science (ARIS2014). J Phys Soc Jpn. https://doi.org/10.7566/aris2014
10. Blumenfeld Y, Nilsson T, Van Duppen P (Jan 2013) Facilities and methods for radioactive ion beam production. Physica Scripta T152:014023. https://doi.org/10.1088/0031-8949/2013/T152/014023
11. Borge MJG, Blaum K (2018) Focus on exotic beams at ISOLDE: a laboratory portrait. J Phys G: Nucl Particle Phys 45(1):010301. https://doi.org/10.1088/1361-6471/aa990f
12. CERN. ISOLDE low energy beamlines. http://cds.cern.ch/record/1515014
13. Symons TJM, Viyogi YP, Westfall GD, Doll P, Greiner DE, Faraggi H, Lindstrom PJ, Scott DK, Crawford HJ, McParland C (1979) Observation of new neutron-rich isotopes by fragmentation of 205-MeV/nucleon 40Ar ions. Phys Rev Lett 42(1):40–43. https://doi.org/10.1103/PhysRevLett.42.40
14. Tanihata I, Hamagaki H, Hashimoto O, Nagamiya S, Shida Y, Yoshikawa N, Yamakawa O, Sugimoto K, Kobayashi T, Greiner DE, Takahashi N, Nojiri Y (1985) Measurements of interaction cross sections and radii of He isotopes. Phys Lett B 160(6):380–384. http://linkinghub.elsevier.com/retrieve/pii/037026938590005X
15. Ärje J, Äystö J, Hyvönen H, Taskinen P, Koponen V, Honkanen J, Valli K, Hautojärvi A, Vierinen K (1986) The ion guide isotope separator on-line, IGISOL. Nucl Instrum Methods Phys Res Sect A: Acceler, Spectrom, Detect Assoc Equipm 247(3):431–437. https://doi.org/10.1016/0168-9002(86)90404-3
16. Facina M, Bruyneel B, Dean S, Gentens J, Huyse M, Kudryavtsev Y, Van den Bergh P, Van Duppen P (2004) A gas cell for thermalizing, storing and transporting radioactive ions and atoms. Part II: On-line studies with a laser ion source. Nucl Instrum Methods Phys Res Sect B: Beam Inter Mater Atoms 226(3):401–418. https://doi.org/10.1016/j.nimb.2004.06.031

17. Hirayama Y, Watanabe Y, Imai N, Ishiyama H, Jeong S, Miyatake H, Oyaizu M, Kim YH, Mukai M, Matsuo Y, Sonoda T, Wada M, Huyse M, Kudryavtsev Y, Van Duppen P (2013) Off-line test of the KISS gas cell. Nucl Instrum Methods Phys Res Sect B: Beam Inter Mater Atoms 317:480–483. https://doi.org/10.1016/j.nimb.2013.06.032
18. Allan Bromley D (ed) (1989) Treatise on heavy-ion science, volume 8, nuclei far from stability. https://doi.org/10.1007/978-1-4613-0713-6
19. Al-Khalili J, Roeckl E (eds) The Euroschool lectures on physics with exotic beams, Vol. I. Lecture notes in physics, vol 651. Springer, Berlin. https://doi.org/10.1007/b98790
20. Al-Khalili J, Roeckl E (eds) (2006) The Euroschool lectures on physics with exotic beams, Vol. II. Lecture notes in physics, vol 700. Springer, Berlin. https://doi.org/10.1007/b11743651
21. Al-Khalili JS, Roeckl E (2009) The Euroschool lectures on physics with exotic beams, Vol. III. Lecture notes in physics, vol 764. Springer, Berlin. https://doi.org/10.1007/978-3-540-85839-3
22. Scheidenberger C, Pfützner M (eds) (2014) The Euroschool on exotic beams, vol. IV. Lecture notes in physics, vol 879. Springer, Berlin. https://doi.org/10.1007/978-3-642-45141-6
23. Borge M, Blaum K (eds) (2017) Focus on exotic beams at ISOLDE: a laboratory portrait. J Phys G: Nucl Particle Phys. http://iopscience.iop.org/journal/0954-3899/page/ISOLDElaboratoryportrait
24. Fedosseev V, Chrysalidis K, Day Goodacre T, Marsh B, Rothe S, Seiffert C, Wendt K (2017) Ion beam production and study of radioactive isotopes with the laser ion source at ISOLDE. J Phys G: Nuclear Particle Phys 44(8):084006. https://doi.org/10.1088/1361-6471/aa78e0
25. Cocolios TE, Al Suradi HH, Billowes J, Budinčević I, de Groote RP, De Schepper S, Fedosseev VN, Flanagan KT, Franchoo S, Garcia Ruiz RF, Heylen H, Le Blanc F, Lynch KM, Marsh BA, Mason PJR, Neyens G, Papuga J, Procter TJ, Rajabali MM, Rossel RE, Rothe S, Simpson GS, Smith AJ, Strashnov I, Stroke HH, Verney D, Walker PM, Wendt KDA, Wood RT (2013) The collinear resonance ionization spectroscopy (CRIS) experimental setup at CERN-ISOLDE. Nucl Instrum Methods Phys Res Sect B: Beam Interact Mater Atoms 317:565–569. https://doi.org/10.1016/j.nimb.2013.05.088
26. Fricke G, Heilig K (2005) 80-Hg mercury. In: Nuclear charge radii. Springer, Berlin, pp 1–9. https://doi.org/10.1007/10856314_82
27. Otten EW (1989) Nuclear radii and moments of unstable isotopes. In: Treatise on heavy ion science. Springer, Boston, pp 517–638. https://doi.org/10.1007/978-1-4613-0713-6_7
28. Ulm G, Bhattacherjee SK, Dabkiewicz P, Huber G, Kluge H-J, Kühl T, Lochmann H, Otten EW, Wendt K, Ahmad SA, Klempt W, Neugart R (1986) Isotope shift of 182Hg and an update of nuclear moments and charge radii in the isotope range 181Hg-206Hg. Zeitschrift für Physik A Hadrons and nuclei 325(3):247–259. https://doi.org/10.1007/BF01294605
29. Zuber K (1935) Separation of the mercury isotopes by a photochemical method. Nature 136(3446):796. https://doi.org/10.1038/136796a0
30. Mrozowski S (1932) Bemerkungen über die Leuchtdauer der Quecksilberresonanzlinie 2537 Å. Z Phys 78(11–12):844–846. https://doi.org/10.1007/BF01342045
31. Melissinos AC (1959) Determination of the dipole moment and isotope shift of radioactive 197Hg by double resonance. Phys Rev 115(1):126–129. https://doi.org/10.1103/PhysRev.115.126
32. Sumner DP, Aung T, Kleiman H (1966) Isotope shifts of neutron-deficient mercury isotopes. Phys Rev 147(3):861–866. https://doi.org/10.1103/PhysRev.147.861
33. Schweitzer Walter G (1963) Hyperfine structure and isotope shifts in the 2537-Å line of mercury by a new interferometric method. J Opt Soc Am 53(9):1055. https://doi.org/10.1364/JOSA.53.001055
34. Bonn J, Huber G, Kluge H-J, Kugler L, Otten EW (1972) Sudden change in the nuclear charge distribution of very light mercury isotopes. Phys Lett B 38(5):308–311. https://doi.org/10.1016/0370-2693(72)90253-5
35. Bonn J, Huber G, Kluge HJ, Köpf U, Kugler L, Otten E-W (1971) Optical pumping of neutron deficient 187Hg. Phys Lett B 36(1):41–43. https://doi.org/10.1016/0370-2693(71)90315-7

36. Hornshøj P, Hansen PG, Jonson B, Lindahl A, Nielsen OB (Mar 1973) Search for rotational fine structure in the nucleus 184Hg. Phys Lett B 43(5):377–379. https://doi.org/10.1016/0370-2693(73)90376-6
37. Nilsson SG, Nix JR, Möller P, Ragnarsson I (1974) How much of a bubble is there in 184Hg? Nucl Phys A 222(2):221–234. https://doi.org/10.1016/0375-9474(74)90389-3
38. Faessler A, Götz U, Slavov B, Ledergerber T (1972) Strongly deformed nuclei near the magic proton shell 82? Phys Lett B 39(5):579–582. https://doi.org/10.1016/0370-2693(72)90001-9
39. Dickmann F, Dietrich K (1973) Variable collective inertia and the transition from spherical to deformed shapes in the Hg-isotopes. Zeitschrift für Physik A Hadrons and nuclei 263(3):211–225. https://doi.org/10.1007/BF01392563
40. Dickmann F, Dietrich K (1974) Coexistence and mixing of spherical and deformed states in the region of light Hg-isotopes. Z Phys 271(4):417–426. https://doi.org/10.1007/BF02126197
41. Cailliau M, Leterssier J, Flocard H, Quentin P (1973) Self consistent calculations of a possible shape transition in neutron deficient mercury isotopes. Phys Lett B 46(1):11–14. https://doi.org/10.1016/0370-2693(73)90461-9
42. Kolb D, Wong CY (1975) Shape isomerism in mercury isotopes. Nucl Phys A 245(2):205–220. https://doi.org/10.1016/0375-9474(75)90173-6
43. Frauendorf S, Pashkkevich VV (1975) On oblate-prolate transition in the ground state rotational band of light mercury isotopes. Phys Lett B 55(4):365–368. https://doi.org/10.1016/0370-2693(75)90360-3
44. Proetel D, Diamond RM, Kienle P, Leigh JR, Maier KH, Stephens FS (1973) Evidence for strongly deformed shapes in 186Hg. Phys Rev Lett 31(14):896–898. https://doi.org/10.1103/PhysRevLett.31.896
45. Rud N, Ward D, Andrews HR, Graham RL, Geiger JS (1973) Lifetimes in the ground-state band of 184Hg. Phys Rev Lett 31(23):1421–1423. https://doi.org/10.1103/PhysRevLett.31.1421
46. Kühl T, Dabkiewicz P, Duke C, Fischer H, Kluge HJ, Kremmling H, Otten EW (1977) Nuclear shape staggering in very neutron-deficient Hg isotopes detected by laser spectroscopy. Phys Rev Lett 39(4):180–183. https://doi.org/10.1103/PhysRevLett.39.180
47. Brack M, Damgaard J, Jensen AS, Pauli HC, Strutinsky VM, Wong CY (1972) Funny hills: the shell-correction approach to nuclear shell effects and its applications to the fission process. Rev Mod Phys 44(2):320–405. https://doi.org/10.1103/RevModPhys.44.320
48. Huber G, Bonn J, Kluge HJ, Otten EW (1976) Nuclear radiation detected optical pumping of neutron-deficient Hg isotopes. Zeitschrift für Physikr Physik A Atoms and Nuclei 276(3):187–202. https://doi.org/10.1007/BF01412097
49. Duke C, Fischer H, Kluge H-J, Kremmling H, Kühl Th, Otten E-W (1977) Determination of the isotope shift of 190Hg by on-line laser spectroscopy. Phys Lett B 68(2):129–132. https://doi.org/10.1016/0370-2693(77)90183-6
50. Dabkiewicz P, Buchinger F, Fischer H, Kluge H-J, Kremmling H, Kühl T, Müller AC, Schuessler HA (1979) Nuclear shape isomerism in 185Hg detected by laser spectroscopy. Phys Lett B 82(2):199–203. https://doi.org/10.1016/0370-2693(79)90735-4
51. Wood JL, Heyde K (2016) A focus on shape coexistence in nuclei. J Phys G: Nucl Part Phys 43(2). https://doi.org/10.1088/0954-3899/43/2/020402
52. Heyde K, Wood JL (2016) Nuclear shapes: from earliest ideas to multiple shape coexisting structures. Phys Scr 91(8). https://doi.org/10.1088/0031-8949/91/8/083008
53. Julin R, Helariutta K, Muikku M (2001) Intruder states in very neutron-deficient Hg, Pb and Po nuclei. J Phys G: Nucl Part Phys 27(7):R109–R139. https://doi.org/10.1088/0954-3899/27/7/201
54. Kondev FG, Carpenter MP, Janssens RVF, Lister CJ, Abu Saleem K, Ahmad I, Amro H, Caggiano J, Davids CN, Heinz A, Herskind B, Khoo TL, Lauritsen T, Ma WC, Ressler JJ, Reviol W, Riedinger LL, Sarantites DG, Seweryniak D, Siem S, Sonzogni AA, Varmette PG, Wiedenhöver I (2002) First observation of excited structures in neutron-deficient 179Hg: evidence for multiple shape coexistence. Phys Lett B, 528(3-4):221–227. https://doi.org/10.1016/S0370-2693(02)01221-2

55. Melerangi A, Appelbe D, Page RD, Boardman HJ, Greenlees PT, Jones P, Joss DT, Julin R, Juutinen S, Kettunen H, Kuusiniemi P, Leino M, Muikku MH, Nieminen P, Pakarinen J, Rahkila P, Simpson J (2003) Shape isomerism and spectroscopy of 177Hg. Phys Rev C 68(4). https://doi.org/10.1103/PhysRevC.68.041301
56. Jenkins DG, Andreyev AN, Page RD, Carpenter MP, Janssens RVF, Lister CJ, Kondev FG, Enqvist T, Greenlees PT, Jones PM, Julin R, Juutinen S, Kettunen H, Kuusiniemi P, Leino M, Leppännen A-P, Nieminen P, Pakarinen J, Rahkila P, Uusitalo J, O'Leary CD, Raddon P, Simons A, Wadsworth R, Joss DT (2002) Confirmation of triple shape coexistence in 179Hg. Phys Rev C 66(1). https://doi.org/10.1103/PhysRevC.66.011301
57. García-Ramos JE, Heyde K (2014) Nuclear shape coexistence: a study of the even-even Hg isotopes using the interacting boson model with configuration mixing. Phys Rev C 89(1). https://doi.org/10.1103/PhysRevC.89.014306
58. Möller P, Sierk AJ, Bengtsson R, Sagawa H, Ichikawa T (2012) Nuclear shape isomers. At Data Nucl Data Tables 98(2):149–300. https://doi.org/10.1016/j.adt.2010.09.002
59. Gaffney LP, Day Goodacre T, Andreyev A, Seliverstov M, Althubiti N, Andel B, Antalic S, Atanasov D, Barzakh AE, Billowes J, Blaum K, Cocolios TE, Cubiss J, Farooq-Smith G, Fedosseev VN, Fedorov DV, Ferrer R, Flanagan KT, Ghys L, Gottberg A, Granados C, Herfurth F, Huyse M, Jenkins DG, Kisler D, Kreim S, Kron T, Kudryavtsev Y, Lunney D, Lynch KM, Manea V, Marsh BA, Mendonca TM, Molkanov PL, Neidherr D, Raabe R, Raeder S, Ramos JP, Rapisarda E, Rosenbusch M, Rossel RE, Rothe S, Schweikhard L, Sels S, Stora T, Tsekhanovich I, Van Beveren C, Van Duppen P, Veinhard M, Wadsworth R, Wienholtz F, Welker A, Wendt K, Wilson GL, Witkins S, Wolf R, Zuber K (2014) Source laser spectroscopy of mercury isotopes. CERN INTC, p 424. https://doi.org/10.17181/CERN.DUGM.L69R

Chapter 2
A Theoretical Understanding of Nuclear Structure

Abstract This chapter gives a brief introduction to the nuclear theory discussed in the rest of the Thesis and some background to shape coexistence in neutron deficient mercury isotopes.

2.1 Introduction

Nuclear models seek to explain and predict the structure and behaviour of nuclei. For heavier nuclei where ab initio calculations are not currently possible due to the computing power required, this is attempted via collective or single particle nuclear models and hybrids of the two. Collective models such as the liquid drop model focus on the structure and symmetries of the macroscopic nuclear system. Refinements and updates of the liquid drop model have been implemented in order to better reproduce specific behaviour observed experimentally. Single particle models such as the shell model are well suited to describing nuclei near shell closures and for nuclei below $A \approx 80$ [1]. Beyond this, the number of valence nucleons makes single particle model calculations challenging. A truncation of the shell model space is required in order to enable such calculations, this technique is employed in models such as the interacting boson model (IBM).

This chapter is largely based on the textbooks of Krane [2], Casten [3], Martin [4] and Povh et al. [5] and the review article of Heyde and Wood [1]. Additional sources are referenced within the text.

2.2 Liquid Drop Type Models

The liquid drop model is a macroscopic model that takes a global view of the nucleus, modelling it as a quantum fluid composed of nucleons that interact with each other resulting in collective effects. Considering the nuclear force, there are a number of

T. Day Goodacre, *Applied Laser Spectroscopy for Nuclear Physics*, Springer Theses,
https://doi.org/10.1007/978-3-030-73889-1_2

similarities between the nucleus and a liquid drop. The volume of a liquid drop is proportional to the number of its constituent molecules and there are no long range interactions between the molecules, thus each molecule interacts only with its neighbours. The liquid drop model was first proposed by Gamow [6], with the nuclear radius R equal to

$$R = R_0 A^{1/3}, \tag{2.1}$$

where R_0 is a constant ≈ 1.2 fm and A is the number of nucleons. It was further developed by von Weizsacker [7] in the 1930s, resulting in the semi empirical mass formula. This models the binding energy $B.E.$ of a nucleus and considers five contributions

$$B.E. = a_V A - a_S A^{2/3} - a_C \frac{Z^2}{A^{1/3}} - a_A \frac{(A-2Z)^2}{A} \pm \delta(A, Z). \tag{2.2}$$

The volume term ($a_V A$) and the surface term ($a_S A^{2/3}$) take into account the effect of the strong nuclear force within the nucleus and at the surface of the nucleus respectively, the Coulomb term ($a_C \frac{Z^2}{A^{1/3}}$) accounts for proton-proton repulsion, the asymmetry term ($a_A \frac{(A-2Z)^2}{A}$) takes into account the Pauli exclusion principle and favours a more equal number of protons and neutrons, finally the pairing term ($\delta(A, Z)$) accounts for the pairing of like nucleons. The volume term contributes towards the binding energy, while the surface, Coulomb and asymmetry terms contribute towards a reduction in the total binding energy. The pairing term can increase or decrease the binding energy for even-even and odd-odd nuclei respectively.

The strength of the liquid drop model is its ability to reproduce global trends in the binding energy per nucleon for the majority of stable nuclei. However, it is unable to reproduce deviations from the nuclear radius $R \propto A^{\frac{1}{3}}$ and phenomena such as asymmetric fission and nuclear magic numbers.

The liquid drop model was updated by Myers and Swiatecki [8–10] to include non-uniform proton and neutron densities in addition to higher order terms and renamed the droplet model. This model includes an additional term $\bar{\varepsilon}$ made up of a number of empirically determined coefficients, included as

$$R = R_0 A^{1/3}(1 + \bar{\varepsilon}). \tag{2.3}$$

The empirical coefficients were updated by D. Berdichevsky and F. Tondeur in 1985, to better agree with experimentally determined isotope shifts [11]. Further updates resulted in the finite range liquid drop model FRLDM and the finite range droplet model FRDM, which have themselves been further refined [12]. These later models represent a macroscopic-microscopic approach, well suited to the modelling of deformed nuclear ground states in the heavier region of the nuclear chart.

2.3 The Shell Model

The shell model considers independent nucleons in an attractive potential, which can be deduced from experimental data. The Woods-Saxon potential $V_{WS}(r)$ is commonly used [13] and can be expressed as

$$V_{WS}(r) = \frac{V_0}{1 + e^{(r-r_0)/a}}, \tag{2.4}$$

where V_0, r_0 and a are constants with r_0 related to the nuclear size and a determining the diffuseness at the nuclear surface. The inclusion of a term for the spin orbit-interaction was proposed by Mayer and Jensen [14, 15], resulting in a potential $V(r)$

$$V(r) = \frac{V_0}{1 + e^{(r-r_0)/a}} + V_{ls}(r)(\mathbf{l} \cdot \mathbf{s}), \tag{2.5}$$

where $\mathbf{l}$ and $\mathbf{s}$ are the orbital and spin angular momentum operators for a single nucleon.

Experimental results indicate an increased stability of nuclei with N or Z = 2, 8, 20, 28, 50, 82, 126 the "magic numbers". These magic numbers are reproduced as successive shell gaps by solving the many-body Schrödinger equation, including the potential $V_{ls}(r)$ [16]. The shell model considers filled shells as an inert core, with the properties of the nucleus determined by the unpaired valence nucleons. The consequence of considering independent particles in the manner of the shell model is that collective excitations are neglected. Using a modified Woods-Saxon or Nilsson potential, the deformed shell model considers independent particles moving in a deformed potential and enables the interpretation of collective excitations [17, 18].

2.4 The Interacting Boson Model (IBM)

The IBM, also known as the interacting boson approximation (IBA), is based on the shell model and was developed to model nuclei far from shell closures [19, 20]. The IBM achieves this by truncating the shell model space, vastly reducing the number of potential configurations.

2.5 Mean Field Theory

In the mean field approach, the nucleons are considered as independent particles interacting with a mean field created by the other nucleons. That is approximated by the Hartree-Fock (Bogoliubov) method and updated at each iteration until a minimal energy is reached. Thus unlike the shell model, there is no inert core. The effective nucleon-nucleon force is tuned to describe the properties of the nuclei and the interactions are included in the Hamiltonian as simplified parametrizations. In relativistic mean field theory the same approach is followed, but the nucleon-nucleon interactions occur via the exchange of mesons.

2.6 Shape Coexistence

Shape coexistence occurs for a given isotope when there is a nucleon configuration (or in the case of ^{186}Pb two [21]) corresponding to a different nuclear shape "coexisting" within a few MeV of the nuclear ground state. The first study of shape coexistence in nuclear physics is identified by [22] to be a 1956 paper by H. Morinaga on the first excited state of ^{16}O [23]. One of the most striking examples is that of the neutron deficient mercury isotopes. A combination of shell effects and residual nucleon-nucleon interactions result in dramatic shifts from a weakly deformed oblate ground state for the even isotopes, to a strongly deformed prolate ground state of the odd isotopes [24, 25]. This shape staggering is observable as pronounced isotope shifts between odd and even isotopes with reducing N from A = 186.

From a shell model interpretation, the shape coexistence in neutron deficient mercury is understood to be the result of an intruder band built on the excitation of two protons (p) from the zero particle two hole (h) proton configuration across the Z = 82 shell gap [26]. The excitation of the two protons is in turn the result of the increasing number of valence neutrons when moving further from the N = 126 neutron shell closure. The competing proton configurations are depicted graphically in Fig. 2.1 for ^{185g}Hg and ^{185m}Hg.

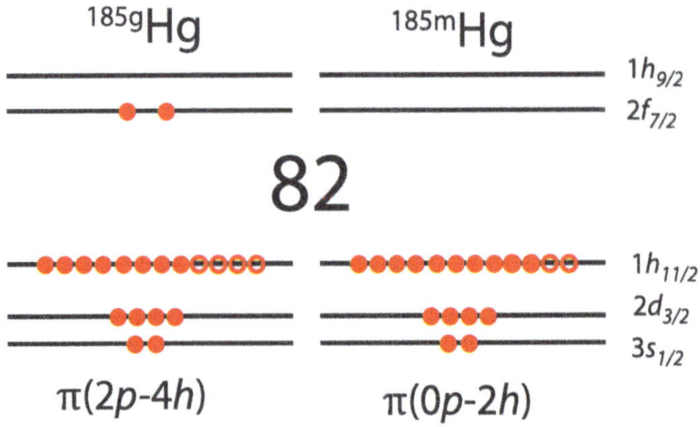

Fig. 2.1 The proton configurations of the ground state and the isomer of ^{185}Hg based on the spherical shell model, level ordering from [3]

Intruder states in the lead region, such as the proton intruder state depicted in Fig. 2.1, can be attributed to the attraction between protons and neutrons. In what could be considered to be the neutron midshell region around N = 104, the numerous valence neutrons are sufficient to excite a pair of protons across the shell gap. In the case of mercury there are therefore two competing proton configurations, one due to the near closed proton shell ($\pi(0p - 2h)$) and the other related to the open neutron shell around N = 104 ($\pi(2p - 4h)$).

References

1. Heyde K, Wood JL (2011) Shape coexistence in atomic nuclei. Rev Mod Phys 83(4):1467–1521. https://doi.org/10.1103/RevModPhys.83.1467
2. Krane KS (1987) Introductory nuclear physics, 3rd edn. Wiley, New York
3. Casten R (2000) Nuclear structure from a simple perspective, 1st edn. Oxford University Press, Oxford
4. Martin BR (2009) Nuclear and particle physics: an introduction, 2nd edn. Wiley, New York
5. Povh B, Rith K, Schulz C, Rodejohann W, Zetsche F (2015) Particles and nuclei: an introduction to the physical concepts, 7th edn. Springer
6. Gamow G (1930) Mass defect curve and nuclear constitution. Proc R Soc A: Math, Phys Eng Sci 126(803):632–644. https://doi.org/10.1098/rspa.1930.0032
7. Weizsacker CFV (1935) Zur Theorie der Kernmassen. Zeitschrift fur Physik 96(7-8):431–458. https://doi.org/10.1007/BF01337700
8. Myers WD, Swiatecki WJ (1969) Average nuclear properties. Ann Phys 55(3):395–505. https://doi.org/10.1016/0003-4916(69)90202-4
9. Myers WD, Swiatecki WJ (1974) The nuclear droplet model for arbitrary shapes. Ann Phys 84(1–2):186–210. https://doi.org/10.1016/0003-4916(74)90299-1
10. Myers WD, Schmidt KH (1983) An update on droplet-model charge distributions. Nucl Phys A 410(1):61–73. https://doi.org/10.1016/0375-9474(83)90401-3
11. Berdichevsky D, Tondeur F (1985) Nuclear core densities, isotope shifts, and the parametrization of the droplet model. Zeitschrift für Physik A Hadrons and nuclei 322(1):141–147. https://doi.org/10.1007/BF01412027
12. Möller P, Sierk AJ, Bengtsson R, Sagawa H, Ichikawa T (2012) Nuclear shape isomers. At Data Nucl Data Tables 98(2):149–300. https://doi.org/10.1016/j.adt.2010.09.002
13. Woods RD, Saxon DS (1954) Diffuse surface optical model for nucleon-nuclei scattering. Phys Rev 95(2):577–578. https://doi.org/10.1103/PhysRev.95.577
14. Haxel O, Jensen JHD, Suess HE (1949) On the "Magic Numbers" in nuclear structure. Phys Rev 75(11):1766–1766. https://doi.org/10.1103/PhysRev.75.1766.2
15. Mayer MG (1949) On closed shells in nuclei II. Phys Rev 75(12):1969–1970. https://doi.org/10.1103/PhysRev.75.1969
16. Mayer MG (1950) Nuclear configurations in the spin-orbit coupling model I. Empirical evidence. Phys Rev 78(1):16–21. https://doi.org/10.1103/PhysRev.78.16
17. Nilsson SG, Tsang CF, Sobiczewski A, Szymański Z, Wycech S, Gustafson C, Lamm IL, Möller P, Nilsson B (1969) On the nuclear structure and stability of heavy and superheavy elements. Nucl Phys A 131(1):1–66. https://doi.org/10.1016/0375-9474(69)90809-4
18. Bengtsson R, Dudek J, Nazarewicz W, Olanders P (1989) A systematic comparison between the Nilsson and Woods-Saxon deformed shell model potentials. Phys Scr 39(2):196–220. https://doi.org/10.1088/0031-8949/39/2/002
19. Arima A, Iachello F (1976) Interacting boson model of collective states I. The vibrational limit. Ann Phys 99(2):253–317. https://doi.org/10.1016/0003-4916(76)90097-X

20. Iachello F, Arima A (1974) Boson symmetries in vibrational nuclei. Phys Lett B 53(4):309–312. https://doi.org/10.1016/0370-2693(74)90389-X
21. Andreyev AN, Huyse M, Van Duppen P, Weissman L, Ackermann D, Gerl J, Hessberger FP, Hofmann S, Kleinböhl A, Münzenberg G, Reshitko S, Schlegel C, Schaffner H, Cagarda P, Matos M, Saro S, Keenan A, Moore C, O'Leary CD, Page RD, Taylor M, Kettunen H, Leino M, Lavrentiev A, Wyss R, Heyde K (2000) A triplet of differently shaped spin-zero states in the atomic nucleus 186Pb. Nature 405(6785):430–433. https://doi.org/10.1038/35013012
22. Wood JL, Heyde K (2016) A focus on shape coexistence in nuclei. J Phys G: Nuclear Particle Phys 43(2):020402. https://doi.org/10.1088/0954-3899/43/2/020402
23. Morinaga H (1956) Interpretation of some of the excited states of 4n self-conjugate nuclei. Phys Rev 101(1):254–258. https://doi.org/10.1103/PhysRev.101.254
24. Bonn J, Huber G, Kluge H-J, Kugler L, Otten EW (1972) Sudden change in the nuclear charge distribution of very light mercury isotopes. Phys Lett B 38(5):308–311. https://doi.org/10.1016/0370-2693(72)90253-5
25. Kühl T, Dabkiewicz P, Duke C, Fischer H, Kluge HJ, Kremmling H, Otten EW (1977) Nuclear shape staggering in very neutron-deficient Hg isotopes detected by laser spectroscopy. Phys Rev Lett 39(4):180–183. https://doi.org/10.1103/PhysRevLett.39.180
26. Gaffney LP, Hackstein M, Page RD, Grahn T, Scheck M, Butler PA, Bertone PF, Bree N, Carroll RJ, Carpenter MP, Chiara CJ, Dewald A, Filmer F, Fransen C, Huyse M, Janssens RVF, Joss DT, Julin R, Kondev FG, Nieminen P, Pakarinen J, Rigby SV, Rother W, Van Duppen P, Watkins HV, Wrzosek-Lipska K, Zhu S (2014) Shape coexistence in neutron-deficient Hg isotopes studied via lifetime measurements in 184,186Hg and two-state mixing calculation. Phys Rev C 89(2):024307. https://doi.org/10.1103/PhysRevC.89.024307

Chapter 3
Nuclear Characteristics in the Optical Spectrum

Abstract This chapter introduces the methodology for studying exotic nuclei via the measurement of atomic transitions which can be driven by laser light. This includes a description of both isotope shifts and the hyperfine structure resulting from magnetic dipole and electric quadrupole moments.

3.1 Introduction

The charge radii and moments of the nucleus affect the energy levels of the atomic electrons, consequently information on the structure of the nucleus can be extracted from the study of these atomic energy levels. One method for such investigations is the study of atomic transitions which can be resonantly driven by laser light.

Unless otherwise stated, the following sections are predominantly based on the textbook of King [1] and the topical reviews of Otten [2], Billowes and Campbell [3], Cheal and Flanagan [4], Cheal et al. [5] and Campbell et al. [6].

3.2 The Isotope Shift

There is a shift in the energy of the atomic fine structure levels between two isotopes of the same element. Considering isotopes A and A', the corresponding shift δ in the transition frequency ν from the atomic ground state can be written as

$$\delta\nu^{A,A'} = \nu^{A'} - \nu^{A}. \tag{3.1}$$

The isotope shift $\delta\nu^{A,A'}$ is a result of the change in the mass of the nuclei (mass shift $\nu_{MS}^{A,A'}$) and the change in the nuclear charge distribution (field shift $\nu_{FS}^{A,A'}$) between the two nuclei. It can therefore be approximated as arising from two distinct components

T. Day Goodacre, *Applied Laser Spectroscopy for Nuclear Physics*, Springer Theses,
https://doi.org/10.1007/978-3-030-73889-1_3

$$\delta\nu^{A,A'} = \nu_{MS}^{A,A'} + \nu_{FS}^{A,A'}. \tag{3.2}$$

The field shift contains the information on the nuclear charge distribution, however, both components of the isotope shift must be evaluated to enable information on the nuclear charge distribution to be extracted.

3.2.1 *The Mass Shift*

The angular momentum of an atomic level is fixed, thus a variation in the mass of a nucleus results in a change in the energy of the atomic energy level. This then corresponds to a variation in the transition frequency ν. $\nu_{MS}^{A,A'}$ is additionally separable into two components. The Normal Mass Shift (NMS) accounts for a change in the reduced mass correction, related to the finite mass of the nucleus and the Specific Mass Shift (SMS) accounts for changes in electron-electron correlations. $\nu_{MS}^{A,A'}$ can then be written as

$$\nu_{MS}^{A,A'} = (N + S)\frac{m^{A'} - m^{A}}{m^{A'}m^{A}}, \tag{3.3}$$

where N and S are constants related to the NMS and SMS respectively, $m^{A'}$ is the mass of isotope A' and m^A is the mass of isotope A. N is calculable, while S requires theoretical models or calibration using non-optical experimental data. From Eq. 3.3, it can be seen that the magnitude of the mass shift has a $1/A^2$ dependence. Consequently, the contribution of $\nu_{MS}^{A,A'}$ to the total isotope shift decreases rapidly with increasing A.

3.2.2 *The Field Shift*

The electron wave function $\psi_e(r)$ overlaps with the nucleus, resulting in a sensitivity to changes in the nuclear charge distribution. Between isotopes of the same element, this manifests as the field shift $\nu_{FS}^{A,A'}$. The evaluation of $\nu_{FS}^{A,A'}$ can therefore enable nuclear structure investigations via the study of atomic energy levels. The field shift can be expressed as

$$\nu_{FS}^{A,A'} = \frac{Ze^2}{6h\epsilon_0}\Delta|\psi_e(0)|^2\delta\langle r^2\rangle^{A,A'}, \tag{3.4}$$

where Z is the number of protons, e is the magnitude of the elementary charge, h is the Planck constant, ϵ_0 is the permittivity of free space, $\Delta|\psi_e(0)|^2$ is the change in the electron density at the nucleus between the two atomic energy levels and $\delta\langle r^2\rangle^{A,A'}$

is the variation in the mean square charge radii between isotopes A and A'. For heavier nuclei, the approximation of a constant $\Delta|\psi_e(0)|^2$ across the nuclear volume becomes less reasonable. Equation 3.4 is then typically replaced by

$$\nu_{FS}^{A,A'} = F\lambda^{A,A'}, \tag{3.5}$$

where F is the electronic factor, a constant that is proportional to the change in electron density at the nucleus between the atomic energy levels and $\lambda^{A,A'}$ is the Seltzer moment [7] which can be approximated as

$$\lambda^{A,A'} = k\langle r^2\rangle^{A,A'}, \tag{3.6}$$

where k is a calculable constant. The variation in charge radii between isotopes A and A' can be compared to the expected variation from a spherical nucleus with an equal volume. To a second order approximation, $\langle r^2\rangle^{A,A'}$ can be expressed as

$$\langle r^2\rangle^{A,A'} = \langle r^2\rangle_0^{A,A'} + \langle r^2\rangle_0 \frac{5}{4\pi}\delta\langle\beta_2^2\rangle^{A,A'}, \tag{3.7}$$

where $\langle r^2\rangle_0$ is the mean square charge radius of the spherical nucleus calculated using Eq. 2.3 and $\delta\langle\beta_2^2\rangle^{A,A'}$ is the change in the mean square quadrupole deformation parameter between the two states. The determination of $\delta\langle\beta_2^2\rangle^{A,A'}$ offers the possibility of a model independent determination of changes in the shape of the nucleus as discussed in Sect. 2.6.

3.3 Isomer Shifts

Isomer shifts are shifts in the atomic fine structure energy level between nuclear ground and isomeric excited states of the same isotope and can be expressed as $\delta\nu^{A^{g,m}}$. In the same manner as described in Sect. 3.2, the study of $\delta\nu^{A^{g,m}}$ enables the comparison of $\langle\beta_2^2\rangle$, which in turn enables the determination of the magnitude of nuclear shape changes between the ground state and the isomer. There is no mass shift component to the isomer shift, this is particularly relevant to the light region of the nuclear chart where the mass shift is dominant.

3.4 Atomic Hyperfine Structure

In addition to isotope shifts of the atomic fine structure levels, the nucleus can also induce a splitting of the atomic fine structure levels themselves, resulting in a hyperfine structure (HFS). The difference in the radius of the nucleus compared to that of the atom, results in only the electric monopole, magnetic dipole and electric quadrupole interactions being relevant to the HFS. The nucleus can only have even moments (M1, E2 etc.), because the nuclear wave function has good parity. Thus it is only those moments that contribute to the HFS. The isotope shift is determined from the substate weighted centroid of the hyperfine multiplet.

3.4.1 The Magnetic Dipole Moment

Nuclei with a non-zero spin (I) have magnetic dipole moment μ_I. The nuclear magnetic moment interacts with the magnetic field of the orbiting electrons, lifting the degeneracy of the magnetic substates of the atomic level (when the total atomic angular momentum $J \neq 0$). The coupling of the atomic angular momentum and nuclear spin means a new quantum number is required. Thus, the states of the hyperfine multiplet are labelled with the total angular momentum quantum number $\boldsymbol{F} = \boldsymbol{I} + \boldsymbol{J}$. F can take values of

$$|I - J| \leqslant F \leqslant |I + J|. \tag{3.8}$$

The magnetic dipole moment shifts the F states from the fine structure energy level by ΔE_A, which can be expressed as

$$\Delta E_A = \frac{A}{2}(F(F+1) - I(I+1) - J(J+1)), \tag{3.9}$$

where A is the hyperfine A factor. Equation 3.9 can be reduced to

$$\Delta E_A = \frac{AK}{2}, \tag{3.10}$$

where $K = (F(F+1) - I(I+1) - J(J+1))$. A is related to μ_I as

$$A = \frac{\mu_I B_J}{IJ}, \tag{3.11}$$

where B_J is the magnetic field produced by the electrons at the site of the nucleus.

3.4.2 The Electric Quadrupole Moment

The contribution of an electric quadrupole moment to the hyperfine atomic structure is the result of an interaction between the average electric field gradient V_{JJ} of the atomic electrons at the site of the nucleus and non-spherical nuclear charge distributions. The electric quadrupole moment results in a perturbation of the HFS ΔE_B, where

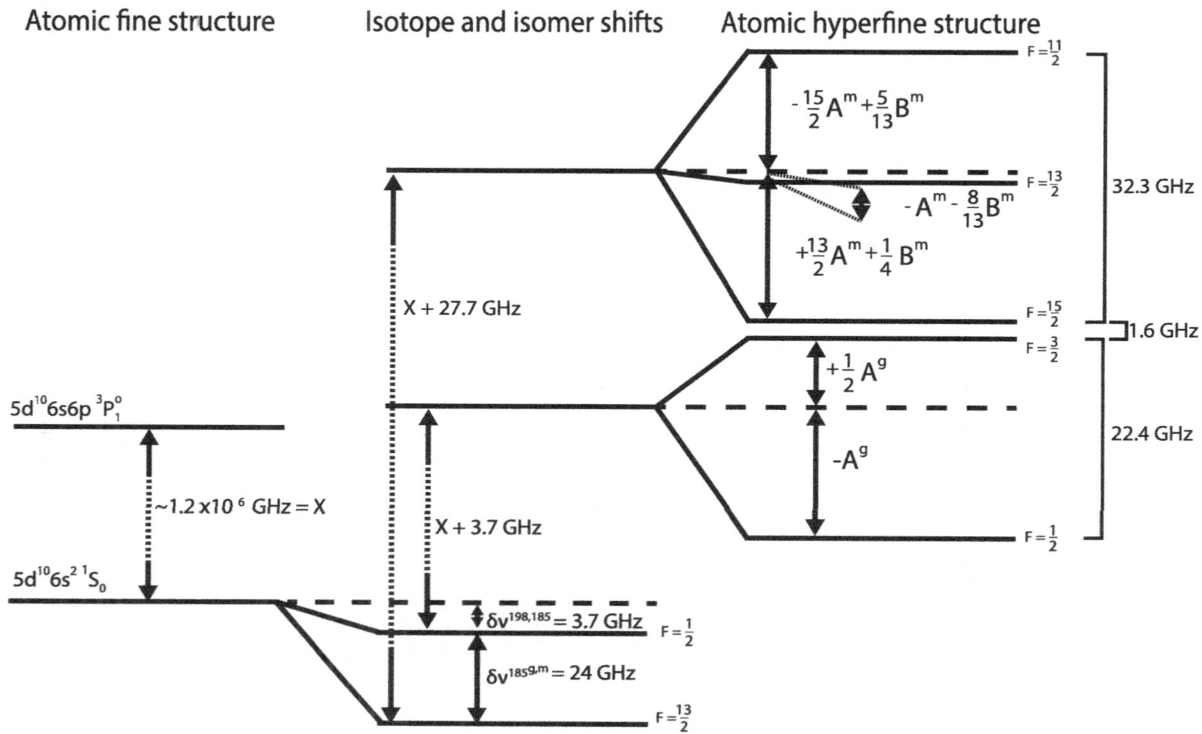

Fig. 3.1 The hyperfine structure of $^{185m}\mathrm{Hg}^{g,m}$, the isotope shift $\delta\nu^{198,185}$ and the isomer shift $\delta\nu^{185g,m}$ are presented. Values for the hyperfine A and B factors were taken from [8] with an A factor of 14.96 GHz for the ground state of the nucleus and hyperfine A and B factors of –2.305 GHz and –0.14 GHz respectively for the isomeric state of the nucleus

$$\Delta E_B = B\frac{\frac{3}{4}K(K+1) - I(I+1)J(J+1)}{2(2I-1)(2J-1)IJ}, \tag{3.12}$$

where B is the hyperfine B factor. B is related to the spectroscopic quadrupole moment Q_s as

$$B = eQ_sV_{JJ}. \tag{3.13}$$

3.4.3 *The Hyperfine Structure in an Atomic Line*

The presence of HFS in an atomic fine structure transition can include contributions from the hyperfine splitting of both the upper and lower atomic energy levels. The total splitting of each level (ΔE_{HFS}) combines Eqs. 3.10 and 3.12, resulting in

$$\Delta E_{HFS} = \Delta E_A + \Delta E_B. \tag{3.14}$$

The expected structure for the 254 nm $5d^{10}6s^2\,{}^1S_0 \rightarrow 5d^{10}6s6p\,{}^3P_1^\circ$ transition in atomic ^{185g,m}Hg is presented in Fig. 3.1.

The isotope shift $\delta\nu^{198,185}$ was unexpectedly small and the isomer shift of $\delta\nu^{185^{gm}}$ was unexpectedly large when first observed [9, 10]. The history of these measurements is discussed in Sect. 1.6. The ground state of atomic mercury has $J = 0$ thus, there is no hyperfine splitting. The hyperfine splitting of the $5d^{10}6s6p\,{}^3P_1^\circ$ excited atomic energy level of ^{185g}Hg is the result of the magnetic dipole moment. There is no contribution due to a spectroscopic quadrupole moment because the nuclear ground state has spin $I = 1/2$. In accordance with Eq. 3.8, the hyperfine splitting results in two unique F levels. The hyperfine splitting of the $5d^{10}6s6p\,{}^3P_1^\circ$ excited state of ^{185m}Hg, is due to both the magnetic dipole and electric quadrupole moments, it has three components as from Eq. 3.8 there are three unique F levels. It can also be observed that the magnetic dipole moment makes the dominant contribution to the hyperfine splitting of the ^{185m}Hg $5d^{10}6s6p\,{}^3P_1^\circ$ excited state.

References

1. King WH (1984) Isotope shifts in atomic spectra. Springer, Boston. https://doi.org/10.1007/978-1-4899-1786-7
2. Otten EW (1989) Nuclear radii and moments of unstable isotopes. In: Treatise on heavy ion science. Springer, Boston, pp 517–638. https://doi.org/10.1007/978-1-4613-0713-6_7
3. Billowes J, Campbell P (1995) High-resolution laser spectroscopy for the study of nuclear sizes and shapes. J Phys G: Nucl Part Phys 21(6):707–739. https://doi.org/10.1088/0954-3899/21/6/003

4. Cheal B, Flanagan KT (2010) Progress in laser spectroscopy at radioactive ion beam facilities. J Phys G: Nucl Particle Phys 37(11):113101. https://doi.org/10.1088/0954-3899/37/11/113101
5. Cheal B, Cocolios TE, Fritzsche S (2012) Laser spectroscopy of radioactive isotopes: role and limitations of accurate isotope-shift calculations. Phys Rev A 86(4):042501. https://doi.org/10.1103/PhysRevA.86.042501
6. Campbell P, Moore ID, Pearson MR (Jan 2016) Laser spectroscopy for nuclear structure physics. Progr Particle Nucl Phys 86:127–180. https://doi.org/10.1016/j.ppnp.2015.09.003
7. Seltzer EC (1969) K X-ray isotope shifts. Phys Rev 188(4):1916–1919. https://doi.org/10.1103/PhysRev.188.1916
8. Ulm G, Bhattacherjee SK, Dabkiewicz P, Huber G, Kluge HJ, Kühl T, Lochmann H, Otten EW, Wendt K, Ahmad SA, Klempt W, Neugart R (1986) Isotope shift of 182Hg and an update of nuclear moments and charge radii in the isotope range 181Hg-206Hg. Zeitschrift für Physik A Hadrons and nuclei 325(3):247–259. https://doi.org/10.1007/BF01294605
9. Bonn J, Huber G, Kluge H-J, Kugler L, Otten EW (1972) Sudden change in the nuclear charge distribution of very light mercury isotopes. Phys Lett B 38(5):308–311. https://doi.org/10.1016/0370-2693(72)90253-5
10. Dabkiewicz P, Buchinger F, Fischer H, Kluge H-J, Kremmling H, Kühl T, Müller AC, Schuessler HA (1979) Nuclear shape isomerism in 185Hg detected by laser spectroscopy. Phys Lett B 82(2):199–203. https://doi.org/10.1016/0370-2693(79)90735-4

Chapter 4
Resonance Ionization Spectroscopy for Nuclear Physics

Abstract This chapter introduces resonance ionization spectroscopy (RIS) and its application for the study and production of radioactive ion beams (RIB).

4.1 Resonance Ionization Spectroscopy

The first multi-step selective photoionization was reported in 1971 by Ambartsumyan (Ambartzumian) et al. [1, 2], using ruby laser pumped tunable dye lasers. Advances in laser technology have enabled the widespread application of the technique, two examples of which are outlined in Sects. 4.2.1 and 4.2.2. Resonance ionization spectroscopy (RIS) is extensively explored in the book of Letokov [3]. Unless otherwise stated, this section is predominantly based on that work.

RIS relies on two principles: for a particular element, the distribution of the atomic energy levels is unique and electronic transitions between energy levels can be induced if the atom is illuminated with photons matching the excitation energy. Photons from multiple lasers are typically used to stepwise excite and then liberate a valence electron from atoms of the element of interest. The selectivity of the ionization process is a direct consequence of the excitation of valence electrons via element-unique atomic transitions.

In order to maximize the ionization efficiency, the atomic transitions should be saturated. The energy flux required to saturate the transition between states a and b by optical illumination (Φ_{ab}) can be expressed as

$$\Phi_{ab} = \Phi_s^{ab} \frac{\Delta\omega_L^{ab}}{\Delta\omega_t^{ab}}, \tag{4.1}$$

where Φ_s^{ab} is the required saturation energy flux of the transition itself, $\Delta\omega_L^{ab}$ is the spectral linewidth of the laser driving the transition and $\Delta\omega_t^{ab}$ is the spectral width of the transition between states a and b in the laser-atom interaction region. $\Delta\omega_t^{ab}$

T. Day Goodacre, *Applied Laser Spectroscopy for Nuclear Physics*, Springer Theses,
https://doi.org/10.1007/978-3-030-73889-1_4

is sensitive to the temperature and pressure of the laser-atom interaction region. A complete spectral coverage of the available atoms $\Delta\omega_L^{ab} \geq \Delta\omega_t^{ab}$ is required in order to maximize the efficiency of a resonant excitation. From Eq. 4.1 however, it also follows that Φ_{ab} increases with increasing laser linewidth. Φ_s^{ab} depends on multiple factors and can be defined as

$$\Phi_s^{ab} = \frac{\hbar\omega_{ab}}{2\sigma_{ab}}, \tag{4.2}$$

where $\hbar$ is the reduced Planck constant, ω_{ab} is the frequency of the laser light and σ_{ab} is the cross-section of the atomic transition. Based on Eq. 4.2, it can be seen that the larger the σ_{ab} of a transition, the less laser power is required to achieve saturation. σ_{ab} is determined by parameters related to both the incident photons and the atomic transition, it can be expressed as

$$\sigma_{ab} = \frac{g_a}{g_b}\frac{\lambda_{ab}^2}{2\pi}\frac{A_{ba}}{\Delta\omega_t^{ab}}, \tag{4.3}$$

where g_a and g_b are the degeneracy of the levels a and b respectively, λ_{ab}^2 is the wavelength of the illuminating laser and A_{ba} is the Einstein A coefficient. Considering atomic transitions in the same environment, variations in σ_{ab} are dominated by changes in A_{ba}, which can differ by up to seven orders of magnitude between reported optical transitions. For optical spectroscopy experiments, the saturation of the spectroscopic transition is avoided as it results in a broadening of the resonance.

The methods of ionization typically applied for RIS are summarized in Fig. 4.1. $\omega1$ represents the laser frequency required to reach the ionization potential (IP) of the element of interest. The combination of $\omega2$ and $\omega3$ represents resonance ionization via an autoionizing state, which involves the excitation of two valence electrons with a combined excitation energy above the first ionization potential of the element. In addition to a radiative decay of this doubly excited atomic state, it is possible for one of the excited electrons to transfer its energy to the other, due to the Coulomb interaction between them, resulting in ionization [4]. Ionization via an autoionizing state is typically the most efficient method of resonance ionization, as the final step is a resonant transition. Consequently there is a greater chance of saturating it with the available laser power. The photo-absorption cross-section for resonant transitions in the "hot cavity" surface ion source environment, referred to here after as the hot cavity and discussed in Sects. 4.2 and 5.2.1, can be up to $10^{-11}\,\text{cm}^{-2}$. This is six to eight orders of magnitude larger than the photo-absorption cross-section for non-resonant ionization to the continuum in this environment, represented as the combination of $\omega2$ and $\omega4$ in Fig. 4.1. Ionization can also be achieved via a resonant excitation to a Rydberg level (represented by the combination of $\omega2$ and $\omega5$ in Fig. 4.1) lying of the order of meVs below the ionization potential. Ionization from Rydberg levels is achieved passively via external factors (X) including: black body radiation from the

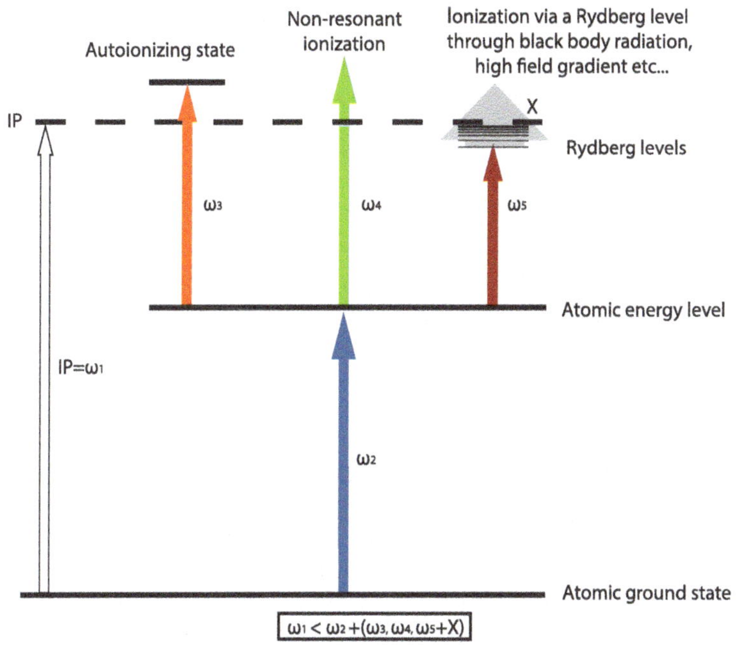

Fig. 4.1 The principle of multi-step resonance photo-ionization. The photon energy required for each of the transitions is depicted as ωx. More than one photon ($\omega 2 - \omega 5$) is required to excite a valence electron above the ionization potential (IP) $= \omega 1$

walls of the hot cavity, the effects of the extraction field gradient or collision with other particles in the atomic vapour. The final laser step in such schemes is a resonant transition, therefore despite the requirement of an additional ionizing contribution, the ionization efficiency can in some cases exceed that of a non-resonant transition to the continuum (depending on the available laser power).

4.2 Radioactive Ion Beam Production

RIS was first proposed as a method of on-line radioactive ion beam production at the "On-line in 1985 and beyond—a workshop on the ISOLDE programme" event [5] in 1984. The technique was demonstrated at the Leningrad Nuclear Physics Institute (now PNPI) in 1988 [6] and at ISOLDE in 1990 [7]. Once established, the experimental set-ups were named Resonance Ionization Laser Ion Sources (RILIS). The technique involves multiple laser beams being directed into a heated metal tube (hot cavity), that is connected to the target where the reaction products are produced. The wider application of RILIS' for radioactive ion beam production was recently summarized in [8], including the combination with gas cells/catchers. This section will only consider hot cavity RILIS (HC-RILIS), as this is the principal method of application discussed in this work (the development of an alternative option is presented in Chap. 6).

4.2.1 Ion Beam Production

The hot cavity, a resistively heated refractory metal tube, provides a defined laser-atom interaction region: confining the reaction products following effusion from the target container via a transfer line. It is typically heated to $\approx$2000 °C, to minimize the wall sticking times of reaction products during their effusion and aid ion survival and extraction. This is discussed in more detail in Sect. 5.2.1. The simplicity of the structure is of particular value for use at radioactive ion beam facilities, where access to the target/ion source for maintenance is typically restricted following on-line installation due to radiation protection concerns. The selectivity of the HC-RILIS can be compromised by competing surface ionization processes (also discussed in Sect. 5.2.1). This can pose a significant problem for certain experiments, particularly if surface ionized contaminants have production yields several orders of magnitude higher than the isotope of interest. To address this, alternative laser-atom interaction regions have been developed for specific cases, these are discussed in Chaps. 5 and 6.

4.2.2 In-Source Resonance Ionization Spectroscopy

In addition offering an element selective method of ionization, the use of resonant atomic transitions results in a sensitivity of the process to the nuclear observables discussed in Chap. 3. Atomic s $\leftrightarrow$ p transitions are typically used as the "spectroscopic transition", because the wave functions of s-electrons have the largest overlap with the nucleus. The efficiency of the ionization process and therefore the ion rate are directly related to the matching of the photon energy of the laser light with the atomic resonances. This enables the isotope shifts and hyperfine splitting of the atomic lines to be investigated.

In-source RIS is typically applied on an atomic vapour in a hot cavity environment, resulting in a limitation of the achievable precision of the measurements due to the Doppler broadening ($\Delta\omega_D$) of the atomic lines. This can be calculated as

$$\Delta\omega_D = 7.16 \times 10^{-7} \omega_0 (T/M)^{1/2}, \tag{4.4}$$

where ω_0 is the frequency of the atomic resonance, T is the temperature and M is the atomic mass [4, 9]. In the hot cavity environment, it is the frequency corresponding to the atomic resonance and the mass that are the dominant factors in variations of $\Delta\omega_D$. At ISOLDE, the RILIS laser wavelength coverage is between 210–950 nm and

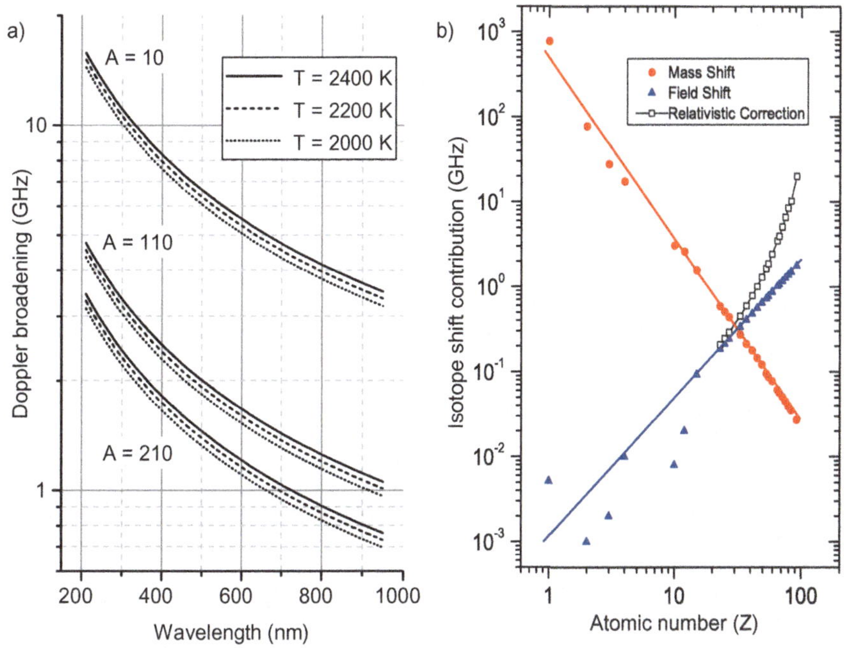

Fig. 4.2 **a** The variation in Doppler broadening for mass (A), temperature (K) and wavelength (nm), calculated from Eq. 4.4, **b** variation in the contribution to the isotope shift with atomic number (Z). Figure (**b**) adapted from [10], © The Royal Swedish Academy of Sciences. Reproduced by permission of IOP Publishing. All rights reserved

its application has ranged from lithium (Z = 3) to radium (Z = 88). By comparison, changes in the temperature of the hot cavity are small: unlikely to vary outside the range of 2000–2400 K. The influence of these parameters is depicted in Fig. 4.2a, together with the contributions to the isotope shift from the mass and field shift (discussed in Sect. 3.2) in Fig. 4.2b.

Based on Fig. 4.2a, b, it is clear that in-source RIS is most applicable in the heavier region of the nuclear chart, where the resonances are less affected by Doppler broadening and the field shift component of the isotope shift dominates the mass shift. The decrease in the mass shift with increasing atomic number can be understood from Eq. 3.3, where the mass shift can be seen to vary as $1/A^2$. The field shift by comparison varies as $Z^2/A^{\frac{1}{3}}$ [10, 11], with a relativistic correction described by [12]. Alternative techniques such as collinear laser spectroscopy or laser spectroscopy on trapped ions (summarised in [13]) offer the possibility of achieving a spectral resolution limited by the natural linewidth of the transition. Therefore when feasible, such techniques offer a significantly higher precision than in-source RIS and are applicable across the nuclear chart. The strength of the in-source RIS technique is its sensitivity. Taking ISOLDE as an example, the isotope with the lowest yield measured by in-source RIS was ^{191}Po, with a rate of 0.01 ions/s on resonance [14], compared to the current reported record for collinear laser spectroscopy of ~100 ions/s [15]. The improving stability, both of the lasers and the experimental set-ups, combined with the sophistication of the detection methods, means that the feasibility of experiments is in many cases becoming limited by the required measurement times.

References

1. Ambartsumyan RV, Kalinin VN, Letokhov VS (1971)Two-step selective photoionization of rubidium atoms by laser radiation. JETP Lett 13(6):217. http://jetpletters.ac.ru/ps/1572/article_24089.shtml
2. Ambartzumian RV, Letokhov VS (1972) Selective two-step (STS) photoionization of atoms and photodissociation of molecules by laser radiation. Appl Opt 11(2):354. https://doi.org/10.1364/AO.11.000354
3. Letokhov VS (1987) Laser photoionization spectroscopy, 1st edn. Academic Press Inc., London
4. Demtröder W (2010) Atoms, molecules and photons. Graduate texts in physics. Springer, Berlin, Heidelberg. http://link.springer.com/10.1007/978-3-642-10298-1, https://doi.org/10.1007/978-3-642-10298-1
5. Letokhov VS, Mishin VI (1984) Laser photoionization pulsed source of radioactive atoms. Book of abstracts, On-line in 1985 and beyond – a workshop on the ISOLDE programme, p D7. http://cds.cern.ch/record/152322
6. Alkhazov GD, Berlovich EYe, Panteleyev VN (1989) A new highly efficient selective laser ion source. Nucl Instrum Methods Phys Res Sect A: Accel, Spectrometers, Detect Assoc Equip 280(1):141–143. https://doi.org/10.1016/0168-9002(89)91282-5
7. Mishin VI, Fedoseyev VN, Kluge H-J, Letokhov VS, Ravn HL, Scheerer F, Shirakabe Y, Sundell S, Tengblad O (1993) Chemically selective laser ion-source for the CERN-ISOLDE on-line mass separator facility. Nucl Instrum Methods Phys Res Sect B: Beam Interact Mater At 73(4):550–560. https://doi.org/10.1016/0168-583X(93)95839-W
8. Marsh BA (2014) Resonance ionization laser ion sources for on-line isotope separators (invited). Rev Sci Instrum 85(2):02B923. https://doi.org/10.1063/1.4858015
9. Fedosseev VN, Fedorov DV, Horn R, Huber G, Köster U, Lassen J, Mishin VI, Seliverstov MD, Weissman L, Wendt K (2003) Atomic spectroscopy studies of short-lived isotopes and nuclear isomer separation with the ISOLDE RILIS. Nucl Instrum Methods Phys Res Sect B: Beam Interact Mater At 204:353–358. https://doi.org/10.1016/S0168-583X(02)01959-6
10. Blaum K, Dilling J, Nörtershäuser W (2013) Precision atomic physics techniques for nuclear physics with radioactive beams. Phys Scr T152:014017. https://doi.org/10.1088/0031-8949/2013/T152/014017
11. Otten EW (1989) Nuclear radii and moments of unstable isotopes. In: Treatise on heavy ion science. Springer, Boston, pp 517–638. https://doi.org/10.1007/978-1-4613-0713-6_7
12. Blundell SA, Baird PEG, Palmer CWP, Stacey DN, Woodgate GK, Zimmermann D (1985) A re-evaluation of isotope shift constants. Z für Phys Phys At Nucl 321(1):31–33. https://doi.org/10.1007/BF01411940
13. Campbell P, Moore ID, Pearson MR (2016) Laser spectroscopy for nuclear structure physics. Prog Part Nucl Phys 86:127–180. https://doi.org/10.1016/j.ppnp.2015.09.003
14. Seliverstov MD, Cocolios TE, Dexters W, Andreyev AN, Antalic S, Barzakh AE, Bastin B, Büscher J, Darby IG, Fedorov DV, Fedoseyev VN, Flanagan KT, Franchoo S, Fritzsche S, Huber G, Huyse M, Keupers M, Köster U, Kudryavtsev Yu, Marsh BA, Molkanov PL, Page RD, Sjødin AM, Stefan I, Van de Walle J, Van Duppen P, Venhart M, Zemlyanoy SG (2013) Charge radii of odd-A $^{191-211}$Po isotopes. Phys Lett B 719(4–5):362–366. https://doi.org/10.1016/j.physletb.2013.01.043

15. Flanagan KT, Lynch KM, Billowes J, Bissell ML, Budinčević I, Cocolios TE, de Groote RP, De Schepper S, Fedosseev VN, Franchoo S, Garcia Ruiz RF, Heylen H, Marsh BA, Neyens G, Procter TJ, Rossel RE, Rothe S, Strashnov I, Stroke HH, Wendt KDA (2013) Collinear resonance ionization spectroscopy of neutron-deficient francium isotopes. Phys Rev Lett 111(21):212501. https://doi.org/10.1103/PhysRevLett.111.212501

Chapter 5
ISOLDE and RILIS

Abstract This chapter introduces the ISOLDE facility and the commonly used ion sources employed there. The resonance ionization laser ion source (RILIS) is presented including its configuration for in-source resonance ionization spectroscopy.

5.1 The ISOLDE Facility

ISOLDE [1] is a world leading radioactive ion beam (RIB) facility based at CERN that has, in various incarnations, been in operation since 1967 [2]. Sadly the TRISTAN radioactive ion beam facility [3], a contemporary of the original ISOLDE facility, is no longer operational. There are a growing number of radioactive ion beam facilities worldwide, offering scientists from disciplines ranging between nuclear structure and nuclear astrophysics to materials and medical sciences the opportunity to conduct experiments with short-lived and exotic isotopes [4]. The planned and currently operating facilities are summarized in [5]. As the name indicates, ISOLDE is an isotope separator on-line, or ISOL type, radioactive ion beam facility. It is classified as a "thick target" ISOL facility, as the target material is of sufficient thickness/density to stop and thermalize the reaction products produced by the impinging "driver beam".

The working principles of ISOLDE are depicted in Fig. 5.1. A pulsed proton beam originating from the LINAC 2 accelerator [6], is further accelerated by the Proton Synchrotron Booster (PSB) [7] to an energy of 1.4 GeV and directed to the ISOLDE facility. The CERN accelerator complex operates a pulsed proton "supercycle" with proton pulses from the PSB separated by 1.2 s. Typically up to 40% of the proton pulses extracted from the PSB are directed to ISOLDE to be impinged on a target.

The majority of the targets used at the ISOLDE facility are uranium carbide (UC_x) as depicted in Fig. 5.1. Reaction products are primarily created by proton driver beam induced fragmentation, spallation or fission of nuclei in the target material. Secondary reactions also take place however, increasing the yield for certain isotopes [8]. The target material is typically heated to between 500 °C and 2200 °C depending on the type. Sufficiently volatile reaction products diffuse out of the target material and then

T. Day Goodacre, *Applied Laser Spectroscopy for Nuclear Physics*, Springer Theses,
https://doi.org/10.1007/978-3-030-73889-1_5

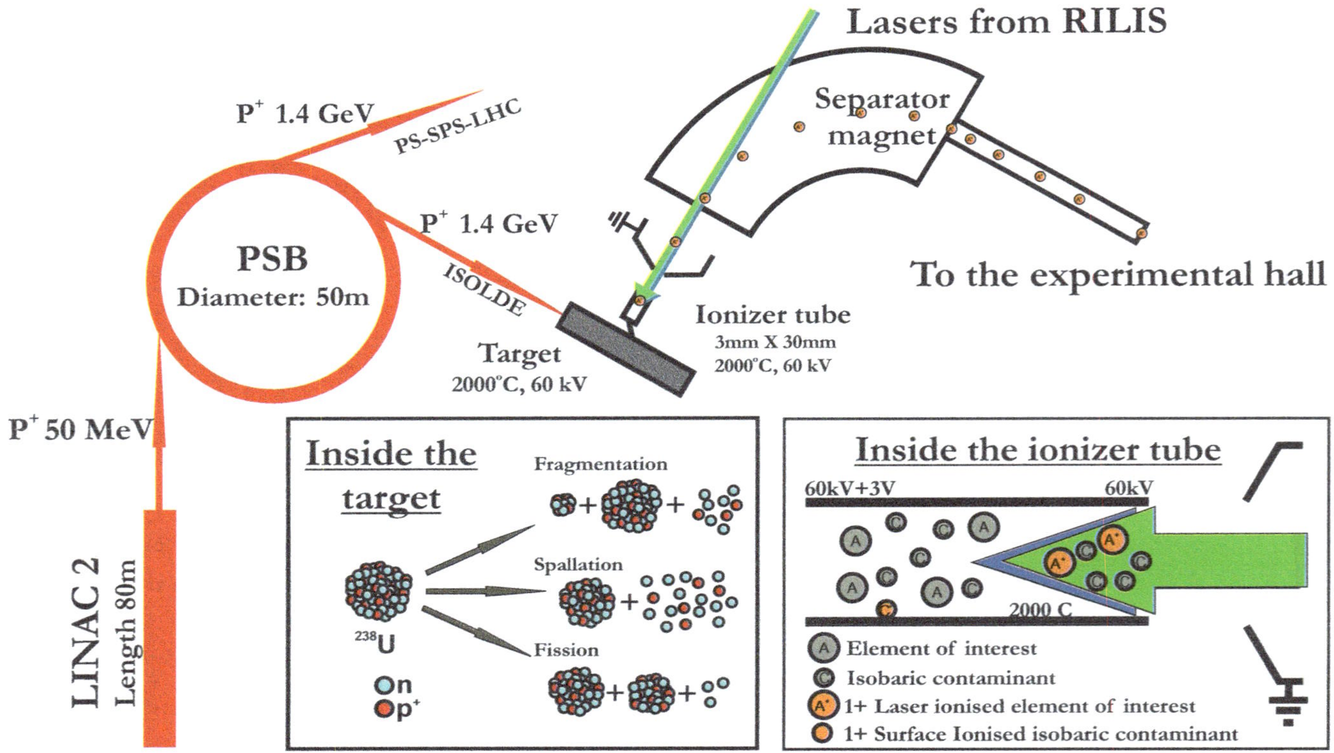

Fig. 5.1 The fundamentals of radioactive ion beam production at ISOLDE. The acceleration stages of the proton driver beam are represented by red arrows. The production and ionization of reaction products is depicted in the inserts. Further discussion in the text

effuse via a transfer line to the ion source, typically for ionization to a 1+ charge state.

The ion source used in the process depicted in Fig. 5.1 is the ISOLDE RILIS, the principle of which is discussed in Sect. 5.2.2. The entire target and ion source assembly is maintained at a voltage between 30 kV–60 kV, with a grounded extraction electrode positioned at the exit of the ion source. The ions are accelerated by the electric field resulting in the extraction of a radioactive ion beam. The ions are then directed through a dipole mass separator, where they are filtered by mv/q (momentum/charge), before being directed to experiments located in the ISOLDE experimental hall.

5.2 ISOLDE Ion Sources

This section will focus on the three most commonly used ion sources at the ISOLDE facility: the hot cavity surface ion source, the RILIS and the VADIS (Versatile Arc Discharge Ion Source). Other less frequently used ion sources, such as negative ion sources [9], the COMIC ion source [10] and the Helicon ion source [11] will not be discussed.

5.2.1 Surface Ion Sources

At ISOLDE, hot cavity surface ion sources [12–14] are used in both their originally envisaged role for surface ionization of sufficiently low ionization potential (IP) elements and as the standard laser-atom interaction region for the ISOLDE-RILIS, this is depicted in Fig. 5.1. The ISOLDE hot cavity surface ion source is a refractory metal tube, 34 mm in length, with an internal diameter of 3 mm, that is typically heated to ≈2000 °C. The heating method is resistive, resulting in a longitudinal potential along the length. This potential accelerates ions created inside the hot cavity towards the extraction aperture.

The efficiency of the surface ion source for a particular element can be estimated using a modified Saha-Langmuir equation, with corrections for the hot cavity environment [15, 16]. An effective IP (IP_{ef}) [17] is used, that takes into account the atomic and ionic excited states. The IP_{ef} can be calculated as

$$IP_{ef} = IP - kT\ln(Q_i/Q_a), \tag{5.1}$$

where k is the Boltzmann constant, T is the temperature in kelvin and Q_i and Q_a are partition functions accounting for the influence of the ground and excited states of the ion or the atom respectively. The partition functions can be calculated as

Table 5.1 Work functions of surface ion source materials determined by thermionic methods for polycrystalline material. Data taken from [22]

Element	Work function (ϕ) (eV)
Tantalum	4.25
Tungsten	4.55
Rhenium	4.72

$$Q_x = \sum_j g_j \exp(-E_j/kT), \tag{5.2}$$

where, g is the degeneracy (2J + 1) of the atomic or ionic state and E_j is the excitation energy of a particular state j. The IP_{ef} can then be inserted into the modified Saha-Langmuir equation to give an expected hot cavity surface ionization efficiency E_{surf}, where

$$E_{surf} = \psi/(1 + \psi), \tag{5.3}$$

$$\psi = N\xi(\exp((\phi - IP_{ef})/kT)), \tag{5.4}$$

and N is the number of collisions with the cavity walls (estimated as the ratio of the surface area of the cavity walls divided by the cross-sectional area of the exit aperture) [15], ξ is the ion survival probability and ϕ is the work function of the cavity surface.

The ion survival probability, ξ, is related to the so-called "hot cavity effect": if the electron emission from the walls is sufficient to dominate the ion load within the cavity, then a plasma potential is understood to build up at the walls, reducing the likelihood of ion-wall collisions and therefore enhancing the ion survival and extraction efficiency [15, 18–20]. This effect is also applicable to laser ions created in the hot cavity surface ion source environment. Sufficient data (RILIS and surface ionization efficiencies) has been obtained experimentally for scandium [21], to enable an estimate to be made of $\xi \approx 0.1$ in a tungsten hot cavity heated to 2400 K.

Inaccuracies in this calculation could stem from several factors. ξ has been demonstrated to be dependent on cavity temperature, surface contamination, ion load and the neutral pressure within the cavity [15], all of which can vary over the course of the operational period. Electron impact ionization has been determined as an additional ionization mechanism, although, ions arising from this process will account for a negligible component of the total ion rate unless $E_{surf} << 1$.

The work functions of the standard ISOLDE hot cavity materials are listed in Table 5.1. From these values it is apparent why, where possible, tantalum is used as the cavity material for the laser ion source, since the lower work function results in a reduced surface ion rate and an increased electron emission at a given temperature.

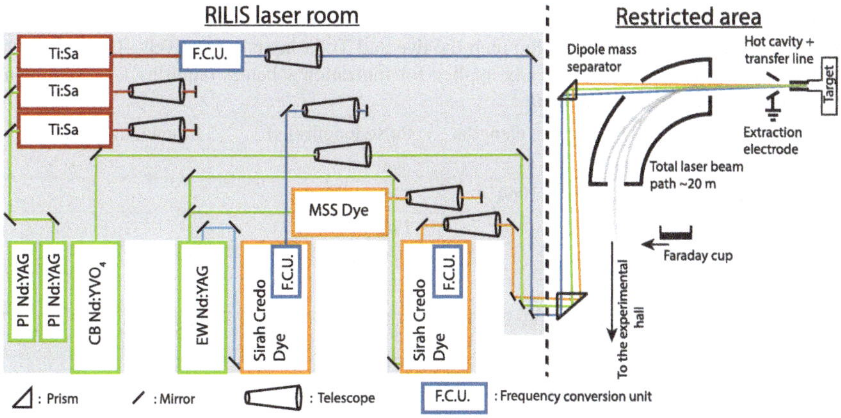

Fig. 5.2 The ISOLDE RILIS, a complete description is given in the text

5.2.2 *The ISOLDE RILIS*

The ISOLDE RILIS [18] has become the principal ion source of the ISOLDE radioactive ion beam facility. It was upgraded in 2012 with the addition of solid state pump lasers and a complementary suite of titanium-doped sapphire (Ti:Sa) lasers [23]. A full description can be found in the latest review article [24]. There are six broadly tunable lasers. This ensures both a good spectral coverage and a sufficient degree of flexibility in the system to meet the requirements of routine operation during the majority of the ISOLDE physics programme. A schematic of the RILIS set-up is presented in Fig. 5.2.

The dye lasers and Ti:Sa lasers are typically pumped by frequency doubled ($2\omega =$ 532 nm) neodymium-doped yttrium aluminium garnet (Nd:YAG) lasers pulsed at 10 kHz. The 10 kHz repetition rate is approximately the frequency required to ensure atoms effusing through the ion source are illuminated by a minimum of one set of laser pulses. Though this is not sufficient for the lightest elements such as beryllium [25]. The pulsed laser light enables sufficient peak intensities for single pass 2ω, 3ω, 4ω frequency conversion, which in combination with the option to pump the dye lasers with the third harmonic of the Nd:YAG ($3\omega = 355$ nm), enables unbroken spectral coverage between 210–950 nm. This range is suitable for the majority of elements other than those such as noble gasses, where the only available transitions from the atomic ground states are for wavelengths <180 nm. An independent 40 W average power, TEM_{00}, frequency doubled ($2\omega = 532$ nm) neodymium-doped yttrium orthovanadate ($Nd:YVO_4$) Lumera Blaze laser, is available for ionization schemes using a non-resonant ionizing transition [26]. An overview of the RILIS laser parameters is presented in Table 5.2, further details can be found in [24].

The laser wavelengths are measured using HighFinesse/Ångstrom WS6 and WS7 wavelength meters. A UV grade fused silica plate, not included in Fig. 5.2, is mounted half way between the laser room and the hot cavity. As the laser beams pass through

Table 5.2 ISOLDE RILIS laser parameters. The EdgeWave (EW) and Photonics Industries (PI) Nd:YAG lasers are typically used to pump the dye and Ti:Sa lasers respectively. The Nd:YVO_4 Coherent Blaze laser (CB) is typically applied for ionization schemes requiring a non-resonant transition to the ionization continuum

Laser	Output wavelengths (nm)	Pulse length (ns)	Linewidth (GHz)
EW Nd:YAG	355, 532, 1064	8	–
PI Nd:YAG	532	170	–
CB Nd:YVO_4	532	17	–
Sirah Credo dye	372–900	7	9
MSS dye	540–900	10	0.8, 15, >60
Ti:Sa (CERN/Mainz)	690–950	40	5, 0.8

this plate, two reference beams are generated (Fresnel reflections from each from each surface). The plate is wedged so that the two reference beams are separated by approximately 20 mm when they enter the reference beam observation area inside the RILIS laboratory. This enables the positions of the laser beams to be stabilized and monitored during the operation of the RILIS. Quadrant photodiodes are used as position sensitive detectors (PSDs), which in combination with piezo-actuated mirrors maintain the laser beam positions.

5.2.3 FEBIAD-Type Ion Sources

A schematic of the principles of FEBIAD-type (Forced Electron Beam Induced Arc Discharge) ion sources [27] is presented in Fig. 5.3. The latest iteration of this ion source at ISOLDE is the VADIS (Versatile Arc Discharge Ion Source) [28]. Electrons are emitted by a cathode heated to ≈2000 °C. The anode is a cylindrical electrode held at 100–150 V, the anode wall facing the cathode is a grid which allows the passage of the accelerated electrons into the ionizer volume [27–29]. A source magnet is used to focus the electrons into helical orbits to increase their path length and therefore the probability of electron-atom collisions inside the anode volume. Reaction products inside the anode volume are ionized by the 100–150 eV electrons from the cathode. Ions are extracted from the 1.5 mm aperture (at the opposite end to the grid) by the penetrating electric field of the ion source casing, that is held at the same voltage as the cathode return (the high voltage platform potential). Similarly to the surface ion source, the entire target unit is maintained at 30–60 kV. The electric field, resulting from a grounded extraction electrode located several centimetres from the exit aperture, extracts and accelerates the ions to create a radioactive ion beam.

The VADIS offers a universal method of ionization, particularly useful for the noble gases. However, because of the universal nature of this ionization process, VADIS ionized radioactive ion beams often suffer from isobaric contamination.

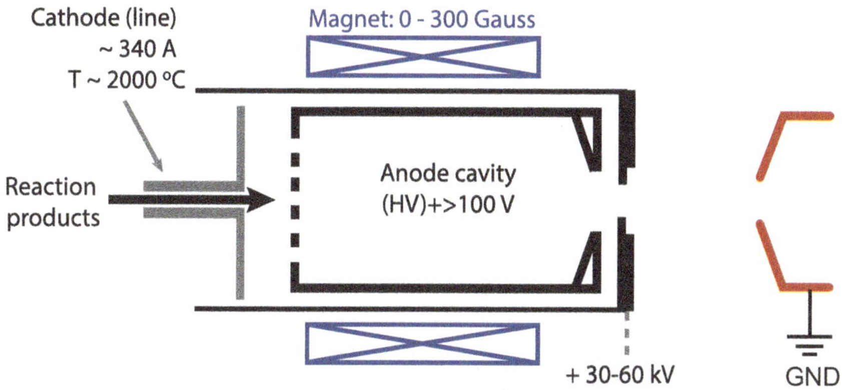

Fig. 5.3 The principle of FEBIAD-type ion sources

Depending on the element of interest, temperature controlled transfer lines can be applied for increased selectivity [29].

5.3 In-Source Spectroscopy at ISOLDE

The experimental set-up for in-source spectroscopy at ISOLDE is detailed in [26]. The laser targeting the spectroscopic transition is scanned in a stepwise manner, while the frequencies of the other transitions are maintained constant. This results in an ion rate of the element of interest that is dependent on the frequency of the laser probing the spectroscopic transition. The resulting ion beam is directed through an ISOLDE dipole mass separator for mass selection and then on to a detection device for the determination of the ion rate at each laser frequency step. Three methods of detection are used. For isotopes with yields greater than 1×10^6 ions/s, an ISOLDE Faraday cup downstream of the dipole mass separator can be used if the isotope of interest is the main component of the mass separated ion beam. For long lived and β emitting isotopes, the ISOLDE Multi-Reflection Time-of-Flight Mass Separator (MR-ToF MS) is used (Sect. 5.3.2). For α-emitters with a sufficiently short half-life, the Windmill detector is used (Sect. 5.3.3).

5.3.1 "Narrowband" RILIS Operation

The fundamentals of the RILIS narrowband ($\approx$1 GHz) Ti:Sa (NB-Ti:Sa) are given in detail in [30]. The wavelength of the laser is scanned using a thin (0.3 mm) and a thick (6 mm) etalon, mounted in motorized mirror mounts and controlled by a LabVIEW program running on the Control PC. A laser scan management and data recording

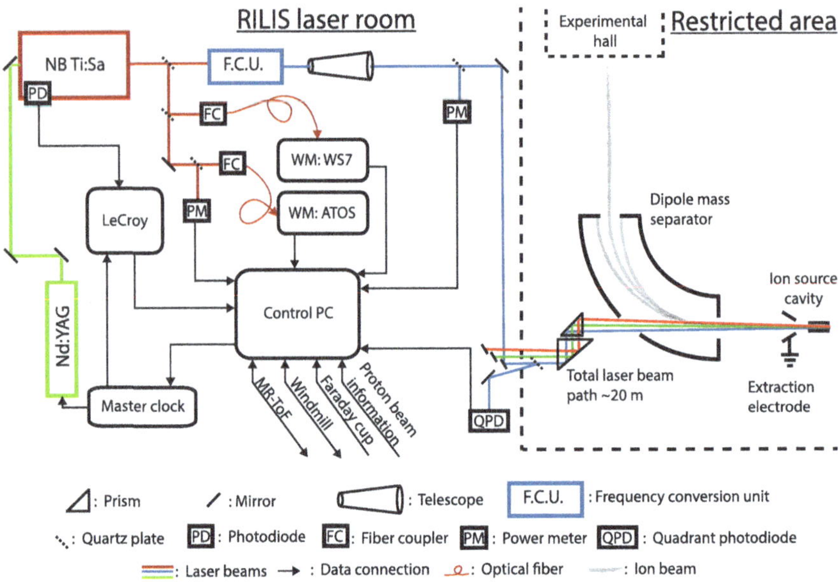

Fig. 5.4 Simplified schematic of the experimental set-up of the scanning laser during in-source RIS experiments at ISOLDE, the direction of the arrows indicates the direction of data flow

program, described in detail in [31], enables the automation of the laser scan: setting the laser frequency step size and coordinating with the Faraday cup, Windmill, or MR-ToF MS detectors. A schematic of the NB-Ti:Sa laser set-up is presented in Fig. 5.4.

The laser light is sampled in multiple locations using UV grade fused silica plates to take two 4% reflections (one from each uncoated surface). The fundamental laser power and the wavelength of the scanning laser are measured before the frequency conversion set-up as depicted in Fig. 5.4. The laser power is then measured a second time on the laser table following frequency conversion and a third time by measuring the power of the reference beam reflected back into the laser room, taking a power reading from the quadrant photodiode PSD used to stabilize the laser beam position. The timing synchronization of the NB-Ti:Sa, with respect to the lasers used for the other excitation steps, is monitored with a photodiode (PD) connected to LeCroy Oscilloscope, that determines the synchronization to nanosecond precision. A LabVIEW program running on the Control PC actively corrects the synchronization of the NB-Ti:Sa during the laser scan. This is achieved by altering the delay on the master clock digital delay generator, which provides the Q-switch trigger for the pump laser.

At each measurement point, the scanning software associates the laser frequency with the ion rate (from either the Faraday cup, the Windmill or the MR-ToF MS), the three laser power measurements, the timing synchronization and the parameters of the proton driver beam. A more refined post analysis of the count rate data is performed

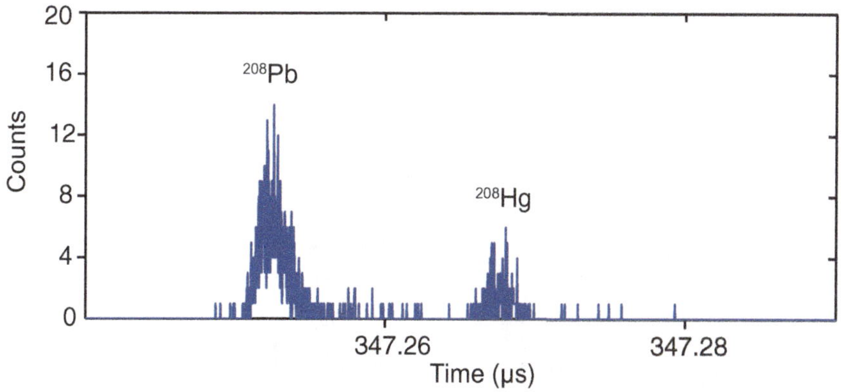

Fig. 5.5 Time-of-flight spectra from a scan of ^{208}Hg. Image courtesy of F. Wienholtz

following the experiment, however, the live information acquired mid-scan enables the quality and stability of the scan to be determined as the scan progresses.

5.3.2 The ISOLTRAP MR-ToF MS

The ISOLTRAP MR-ToF MS [32] enables the selective detection of longer lived and β-emitting isotopes, using a MagneTOF detector installed after a mass spectrometer. The ions are reflected between electrostatic plates $\approx 10^3$ times, separating isobaric contaminants by picosecond differences in the ≈ 30 ms flight time arising from the mass differences between isobars. In most cases, gating on the ion arrival time at the detector enables the determination of the ion rate of the ion beam component of interest, free of isobaric contamination. The use of the MR-ToF MS for in-source RIS is discussed in detail in [33]. A representative time-of-flight mass spectra from one of the measurement points of a scan of ^{208}Hg is presented in Fig. 5.5. The separation of the ^{208}Hg ions from the ^{208}Pb isobaric contamination is clearly apparent.

5.3.3 The Windmill Detector

The Windmill detector consists of a rotating wheel containing 10 thin carbon foils (20 μg/cm^2). The application of the Windmill detector for in-source RIS is described in detail in [34]. The ion beam is deposited onto one of the foils and the α-decay energies are measured by silicon detectors. At the position of deposition, a carbon foil is sandwiched between two silicon detectors: an annular detector to enable the transmission of the ion beam to the carbon foil and a solid detector at the rear. The movement of the wheel is synchronized with the PSB supercycle and the laser scanning the

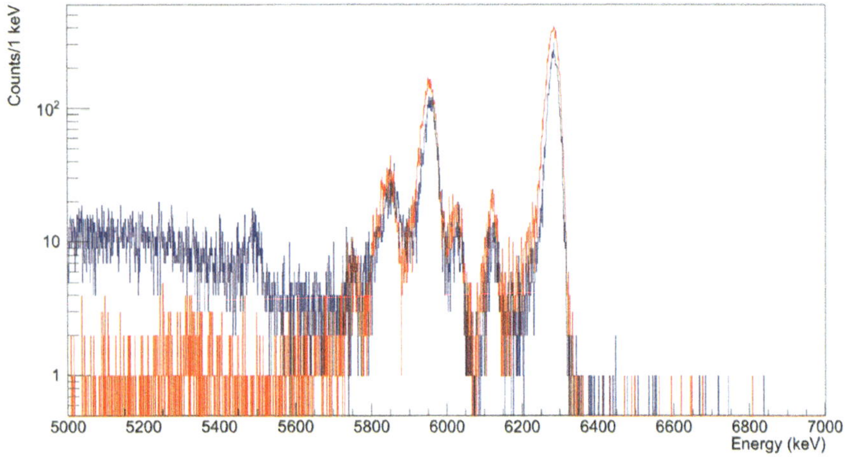

Fig. 5.6 Energy spectra from the measurements of ^{179}Hg. Image courtesy of S. Sels

spectroscopic transition. Gates are set on the α-energies, enabling the determination of the frequency dependent ion rate of the isotope of interest. A representative energy spectra of ^{179}Hg is presented in Fig. 5.6.

References

1. Kugler E (2000) The ISOLDE facility. Hyperfine Interact 129(1/4):23–42. https://doi.org/10.1023/A:1012603025802
2. Allardyce BW (1986) ISOLDE today and tomorrow. In: Richard A Meyer, Daeg S Brenner (eds) Nuclei off the line of stability. ACS symposium series, vol 324, chap 61. American Chemical Society, Washington, DC, pp 406–413. https://doi.org/10.1021/bk-1986-0324.ch061
3. McConnell JR, Talbert WL (1975) The Tristan on-line isotope separator facility. Nucl Instrum Methods 128(2):227–243. https://doi.org/10.1016/0029-554X(75)90672-2
4. Jonson B, Richter A (2000) More than three decades of ISOLDE physics. Hyperfine Interact 129(1/4):1–22. https://doi.org/10.1023/A:1012689128103
5. Blumenfeld Y, Nilsson T, Van Duppen P (2013) Facilities and methods for radioactive ion beam production. Phys Scr 2013(T152):14023. https://doi.org/10.1088/0031-8949/2013/T152/014023
6. Boltezar E, Haseroth H, Pirkl W, Plass G, Sherwood TR, Tallgren U, Tetu P, Warner D, Weiss M (1979) Performance of the new CERN 50 MeV Linac. IEEE Trans Nucl Sci 26(3):3674–3676. https://doi.org/10.1109/TNS.1979.4330576
7. Reich KH (1976) The CERN PS booster, a flexible and reliable injector. In: Proceedings of the 5th All-Union Conference on charged-particle accelerators, pp 221–225, Dubna. https://cds.cern.ch/record/1035486
8. Lukić S, Gevaert F, Kelić A, Ricciardi MV, Schmidt KH, Yordanov O (2006) Systematic comparison of ISOLDE-SC yields with calculated in-target production rates. Nucl Instrum Methods Phys Res Sect A: Accel, Spectrometers, Detect Assoc Equip 565(2):784–800. https://doi.org/10.1016/j.nima.2006.04.082

9. Kirchner R (1981) Progress in ion source development for on-line separators. Nucl Instrum Methods Phys Res 186(1–2):275–293. https://doi.org/10.1016/0029-554X(81)90916-2
10. Sortais P, Lamy T, Médard J, Angot J, Latrasse L, Thuillier T (2010) Ultracompact/ultralow power electron cyclotron resonance ion source for multipurpose applications. Rev Sci Instrum 81(2):02B314. https://doi.org/10.1063/1.3272878
11. Kronberger M, Gottberg A, Mendonca TM, Ramos JP, Seiffert C, Suominen P, Stora T (2013) Production of molecular sideband radioisotope beams at CERN-ISOLDE using a Helicon-type plasma ion source. Nucl Instrum Methods Phys Res Sect B: Beam Interact Mater At 317:438–441. https://doi.org/10.1016/j.nimb.2013.07.032
12. Beyer GJ, Herrmann E, Piotrowski A, Raiko VJ, Tyrroff H (1971) A new method for rare-earth isotope separation. Nucl Instrum Methods 96(3):437–439. https://doi.org/10.1016/0029-554X(71)90613-6
13. Johnson PG, Bolson A, Henderson CM (1973) A high temperature ion source for isotope separators. Nucl Instrum Methods 106(1):83–87. https://doi.org/10.1016/0029-554X(73)90049-9
14. Ravn HL, Sundell S, Westgaard L (1975) Target Techniques for the isolde on-line isotope separator. Nucl Instrum Methods 123(1):131–144. https://doi.org/10.1016/0029-554X(75)90089-0
15. Kirchner R (1990) On the thermoionization in hot cavities. Nucl Instrum Methods Phys Res Sect A: Accel, Spectrometers, Detect Assoc Equip 292(2):203–208. https://doi.org/10.1016/0168-9002(90)90377-I
16. Liu Y, Bilheux HZ, Kawai Y (2005) Characterization of a tubular hot-cavity surface ion source. In: Proceedings of the 21st IEEE particle accelerator conference (PAC 05), 16–20 May. Knoxville, Tennessee, pp 1581–1583. http://accelconf.web.cern.ch/AccelConf/p05/PAPERS/TPPE018.PDF
17. Zandberg ÉY, Ionov NI (1959) Surface ionization. Sov Phys Uspekhi 2(2):255–281. https://doi.org/10.1070/PU1959v002n02ABEH003123
18. Mishin VI, Fedoseyev VN, Kluge H-J, Letokhov VS, Ravn HL, Scheerer F, Shirakabe Y, Sundell S, Tengblad O (1993) Chemically selective laser ion-source for the CERN-ISOLDE on-line mass separator facility. Nucl Instrum Methods Phys Res Sect B: Beam Interact Mater At 73(4):550–560. https://doi.org/10.1016/0168-583X(93)95839-W
19. Lettry J, Catherall R, Focker GJ, Jonsson OC, Kugler E, Ravn H, Tamburella C, The ISOLDE Collaboration, Fedoseyev V, Mishin VI, Huber G, Sebastian V, Koizumi M, Köster U (1998) Recent developments of the ISOLDE laser ion source. Rev Sci Instrum 69(2):761. https://doi.org/10.1063/1.1148978
20. Maden C, Baur H, Fauré A-L, Hubert A, Pointurier F, Bourdon B (2016) Determination of ionization efficiencies of thermal ionization cavity sources by numerical simulation of charged particle trajectories including space charge. Int J Mass Spectrom 405:39–49. https://doi.org/10.1016/j.ijms.2016.05.013
21. Fedosseev VN, Marsh BA, Fedorov DV, Köster U, Tengborn E (2005) Ionization scheme development at the ISOLDE RILIS. In: Laser 2004. Springer, Berlin/Heidelberg, pp 15–27. https://doi.org/10.1007/3-540-30926-8_2
22. CRC (2017) Physical constants of organic compounds, in CRC handbook of chemistry and physics, 97th edn. CRC Press/Taylor & Francis, Boca Raton. http://hbcponline.com/
23. Fedosseev VN, Berg L-E, Fedorov DV, Fink D, Launila OJ, Losito R, Marsh BA, Rossel RE, Rothe S, Seliverstov MD, Sjödin AM, Wendt KDA (2012) Upgrade of the resonance ionization laser ion source at ISOLDE on-line isotope separation facility: new lasers and new ion beams. Rev Sci Instrum 83(2):02A903. https://doi.org/10.1063/1.3662206
24. Rothe S, Day Goodacre T, Fedorov DV, Fedosseev VN, Marsh BA, Molkanov PL, Rossel RE, Seliverstov MD, Veinhard M, Wendt KDA (2016) Laser ion beam production at CERN-ISOLDE: new features more possibilities. Nucl Instrum Methods Phys Res Sect B: Beam Interact Mater At 376:91–96. https://doi.org/10.1016/j.nimb.2016.02.024
25. Köster U, Argentini M, Catherall R, Fedoseyev VN, Gäggeler HW, Jonsson OC, Weinreich R (2003) Off-line production of intense $^{7,10}Be_+$ beams. Nucl Instrum Methods Phys Res Sect B: Beam Interact Mater At 204:343–346. https://doi.org/10.1016/S0168-583X(02)01952-3

26. Marsh BA, Andel B, Andreyev AN, Antalic S, Atanasov D, Barzakh AE, Bastin B, Borgmann C, Capponi L, Cocolios TE, Day Goodacre T, Dehairs M, Derkx X, De Witte H, Fedorov DV, Fedosseev VN, Focker GJ, Fink DA, Flanagan KT, Franchoo S, Ghys L, Huyse M, Imai N, Kalaninova Z, Köster U, Kreim S, Kesteloot N, Kudryavtsev Y, Lane J, Lecesne N, Liberati V, Lunney D, Lynch KM, Manea V, Molkanov PL, Nicol T, Pauwels D, Popescu L, Radulov D, Rapisarda E, Rosenbusch M, Rossel RE, Rothe S, Schweikhard L, Seliverstov MD, Sels S, Sjödin AM, Truesdale V, Van Beveren C, Van Duppen P, Wendt K, Wienholtz F, Wolf RN, Zemlyanoy SG (2013) New developments of the in-source spectroscopy method at RILIS/ISOLDE. Nucl Instrum Methods Phys Res Sect B: Beam Interact Mater At 317:550–556. https://doi.org/10.1016/j.nimb.2013.07.070
27. Kirchner R, Roeckl E (1976) Investigation of gaseous discharge ion sources for isotope separation on-line. Nucl Instrum Methods 133(2):187–204. https://doi.org/10.1016/0029-554X(76)90607-8
28. Penescu L, Catherall R, Lettry J, Stora T (2010) Development of high efficiency versatile arc discharge ion source at CERN ISOLDE. Rev Sci Instrum 81(2):02A906. https://doi.org/10.1063/1.3271245
29. Sundell S, Ravn H (1992) Ion source with combined cathode and transfer line heating. Nucl Instrum Methods Phys Res Sect B: Beam Interact Mater At 70(1–4):160–164. https://doi.org/10.1016/0168-583X(92)95926-I
30. Rothe S, Fedosseev VN, Kron T, Marsh BA, Rossel RE, Wendt KDA (2013) Narrow linewidth operation of the RILIS titanium: sapphire laser at ISOLDE/CERN. Nucl Instrum Methods Phys Res Sect B: Beam Interact Mater At 317:561–564. https://doi.org/10.1016/j.nimb.2013.08.058
31. Rossel RE (2015) A distributed monitoring and control system for the laser ion source RILIS at CERN-ISOLDE. PhD thesis, Hochschule Rhein-Main. https://doi.org/10.17181/CERN.L5N9.2GCS
32. Wolf RN, Beck D, Blaum K, Böhm Ch, Borgmann Ch, Breitenfeldt M, Herfurth F, Herlert A, Kowalska M, Kreim S, Lunney D, Naimi S, Neidherr D, Rosenbusch M, Schweikhard L, Stanja J, Wienholtz F, Zuber K (2012) On-line separation of short-lived nuclei by a multi-reflection time-of-flight device. Nucl Instrum Methods Phys Res Sect A: Accel, Spectrometers, Detect Assoc Equip 686:82–90. https://doi.org/10.1016/j.nima.2012.05.067
33. Wolf RN, Wienholtz F, Atanasov D, Beck D, Blaum K, Borgmann Ch, Herfurth F, Kowalska M, Kreim S, Litvinov YuA, Lunney D, Manea V, Neidherr D, Rosenbusch M, Schweikhard L, Stanja J, Zuber K (2013) ISOLTRAP's multi-reflection time-of-flight mass separator/spectrometer. Int J Mass Spectrom 349–350:123–133. https://doi.org/10.1016/j.ijms.2013.03.020
34. Seliverstov MD, Cocolios TE, Dexters W, Andreyev AN, Antalic S, Barzakh AE, Bastin B, Büscher J, Darby IG, Fedorov DV, Fedosseev VN, Flanagan KT, Franchoo S, Huber G, Huyse M, Keupers M, Köster U, Kudryavtsev Yu, Marsh BA, Molkanov PL, Page RD, Sjödin AM, Stefan I, Van Duppen P, Venhart M, Zemlyanoy SG (2014) Electromagnetic moments of odd-A Po193-203,211 isotopes. Phys Rev C 89(3):034323. https://doi.org/10.1103/PhysRevC.89.034323

Chapter 6
The VADLIS Ion Source

Abstract This chapter describes the development of the versatile arc discharge and laser ion source (VADLIS) which is the combination of the ISOLDE RILIS with ISOLDE's variant of the forced electron beam induced arc discharge (FEBIAD) ion source.

6.1 Introduction

The development of alternative laser-atom interaction regions offers the possibility to increase the applicability of the ISOLDE RILIS. A successful example of this is the LIST (Laser Ion Source and Trap) [1, 2], which has been demonstrated to reduce significantly the surface ionized contamination often present during hot cavity (HC)-RILIS operation. Despite a reduction in yield of one to two orders of magnitude, the selectivity offered by the LIST enabled the successful in-source RIS of ^{217}Po and ^{219}Po [3]. This chapter presents the investigation, characterization and early application of the coupling of the RILIS (described in Sect. 5.2.2) with the VADIS ion source (described in Sect. 5.2.3). This combination that has been termed the VADLIS (Versatile Arc Discharge and Laser Ion Source). Sections 6.2, 6.4 and 6.6 are based upon the publication [4] in Nuclear Instruments and Methods in Physics Research Section B: Beam Interactions with Materials and Atoms titled: "*Blurring the boundaries between ion sources: The application of the RILIS inside a FEBIAD type ion source at ISOLDE*". The paper is licensed under CC by 4.0.

The original motivation for coupling the RILIS and the VADIS was as a step towards the production of refractory metal ion beams at ISOLDE: extracting refractory elements as molecules [5], dissociation to atomic form using the VADIS, then selective ionization using the RILIS lasers. Developments in this direction continue but the initial results for resonance laser ionization of non-refractory elements has led to both immediate applications and additional opportunities.

T. Day Goodacre, *Applied Laser Spectroscopy for Nuclear Physics*, Springer Theses,
https://doi.org/10.1007/978-3-030-73889-1_6

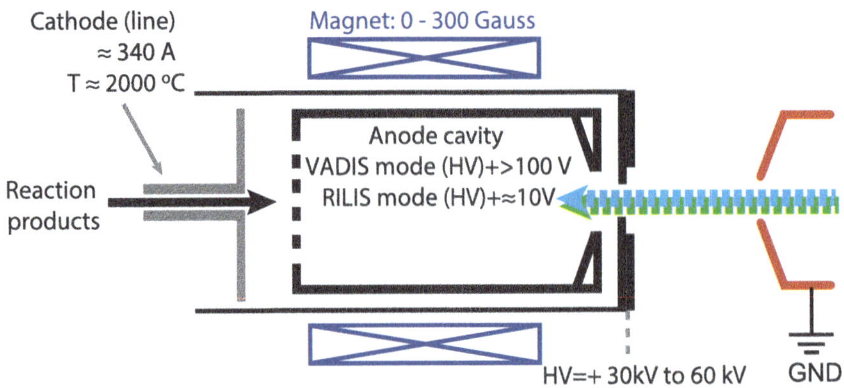

Fig. 6.1 A schematic of the VADLIS ion source. Figure adapted from [4] licensed under CC by 4.0

6.2 Combining the ISOLDE RILIS and VADIS

The anode geometries of all ISOLDE VADIS type ion sources are identical to that of the MK5 FEBIAD [6, 7]. The difference came from replacing the remaining graphite components with molybdenum, due to concerns about the influence of out-gassing leading to significant amounts of CO in the ion source [6, 8] and standardising all variants with the MK5 anode geometry. Consequently, the results presented here are expected to be equally applicable to FEBIAD variants used at radioactive ion beam facilities across the world.

In standard HC-RILIS operation the atoms are confined by the hot cavity, enabling a full overlap of the laser light with the reaction products that have effused into the laser atom interaction region. In the case of the VADLIS, the laser light is transmitted through the 1.5 mm extraction aperture as depicted in Fig. 6.1. The geometry of the anode cavity therefore limits the laser-atom overlap region to <7% of the anode volume. If however the laser excitations steps are saturated, the efficiency of laser resonance ionization within the cavity is independent of the area of the extraction aperture [9, 10]. In principle, the challenge is therefore the efficient extraction of isotopes of interest ionized within the anode cavity.

6.3 Experimental Set-Up at the ISOLDE Off-Line Separator

The first VADLIS tests took place at the ISOLDE off-line separator, which is used for target and ion source development and acceptance tests before installation and operation on-line. The experimental set-up is summarized in Fig. 6.2. The off-line facility is a replica of the ISOLDE on-line front-end and is connected to a dipole mass separator magnet. Gallium was identified as a test element to enable a simplified laser set-up (applying a RILIS ionization scheme of $\{\lambda_1|\lambda_2\}$=$\{294\ \text{nm}|532\ \text{nm}^{NR}\}$) [11]. Laser and surface ionization of gallium has been well characterized previously in the hot cavity [12]. The vapour pressure [13] and the natural isotopic abundance signature of gallium make it a reliable and easily identifiable ion beam for ion source development work. Additionally, the effective ionization potential of 6.3 eV (calculated using Eq. (5.1) and atomic data from [14]) ensures that gallium is surface ionized at a typical hot cavity operating temperature of 2000 °C, but with low efficiency ($<1\%$), thereby providing a benchmark by which to judge the relative efficiencies of other ionization methods. The gallium was supplied through the evaporation of a gallium sample, with a natural isotopic distribution, that was loaded into an independently heated capillary "mass marker" [5] connected upstream of the ion source.

The 294 nm first step was produced by frequency tripling light from a Ti:Sa laser. The non-resonant ionizing transition of 532 nm was provided by a commercial frequency doubled Nd:YAG laser from EdgeWave. The lasers parameters are given in Table 5.2. The wavelength of the Ti:Sa laser was measured using a HighFinesse/Ångstrom WS6 wavelength meter. The laser beams were directed through the window in the dipole mass separator magnet and into the anode cavity of a VADIS via the exit aperture as depicted in Fig. 6.1. The laser parameters (beam position, average power, wavelength, timing) were maintained constant while the VADIS parameters (anode voltage, cathode heating, magnetic field strength) were varied to investigate the optimal conditions for the extraction of laser ions. The target and ion source construction was held at +32 kV, a grounded extraction electrode positioned downstream of the anode aperture extracted ions from the mouth of the ion source and accelerated them to form a 32 keV ion beam.

The time structure of the laser ions was investigated by using a LeCroy Waverunner 104MXi oscilloscope to a monitor the ion current detected with an MCP (microchannel plate) detector inserted in the focal plane of the dipole separator magnet. The oscilloscope was synchronised with the 10 kHz laser trigger to enable time structure histograms to be recorded, plotting the ion arrival time with respect to the incident laser pulses. A Faraday cup was used to measure ion currents in excess of 1 pA providing a more robust measurement device but without the time resolution capabilities of the MCP.

6.4 Off-Line Testing and Characterization

The effect of varying the anode voltage on the ion current is depicted in Fig. 6.3a and Fig. 6.3b. The total current with the lasers applied is shown as dashed blue lines. The ion current with the lasers blocked (termed VADIS ion current) is plotted with a black line and the difference between the two is plotted as a red line and labelled Laser ion current.

The results are interpreted in detail in [4], in addition to the results of the application of the RILIS to ionize barium in a VADIS anode cavity. There is a distinct

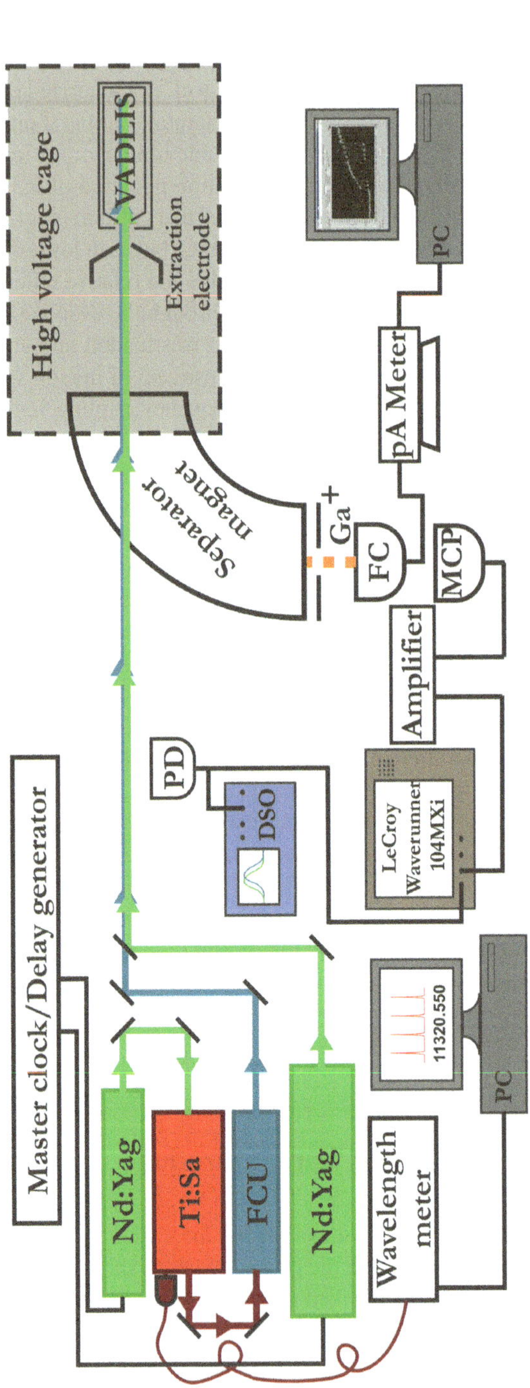

Fig. 6.2 Experimental set-up used at the during tests at the ISOLDE off-line separator. DSO: digital storage oscilloscope, FC: Faraday cup, FCU: frequency conversion unit, MCP: microchannel plate, PD: photo diode. Figure reproduced from [4] licenced under CC by 4.0

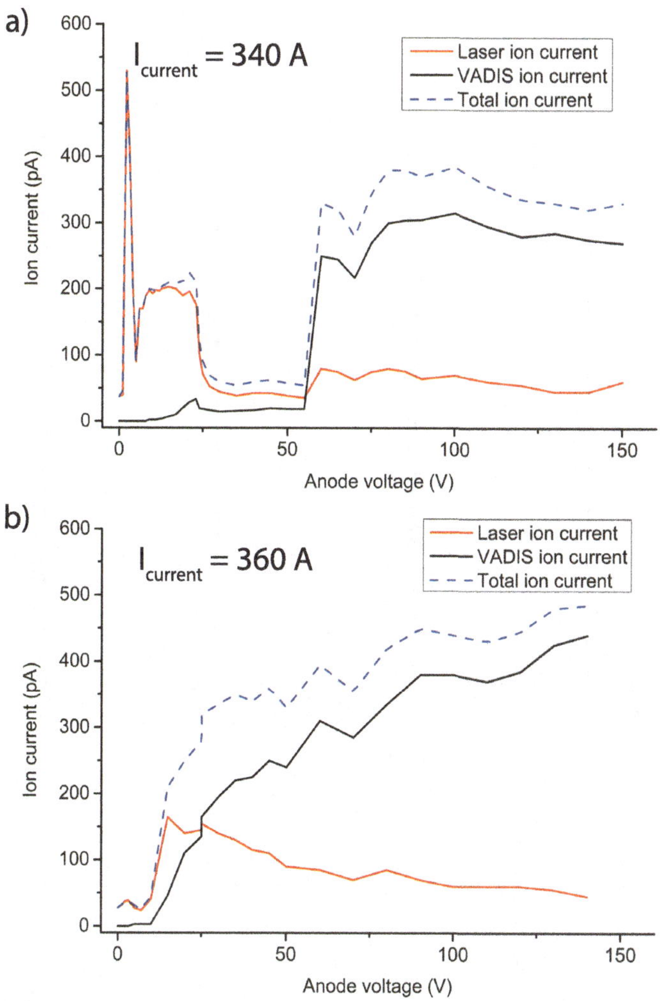

Fig. 6.3 Anode voltage versus ^{71}Ga ion current for an applied cathode heating current of 340 A (**a**) and 360 A (**b**). The laser ion current was determined by measuring the reduction in the total ion current following the blocking of the 294 nm laser beam. Figure reproduced from [4] licensed under CC by 4.0

RILIS mode of operation visible in Fig. 6.3a for anode voltages <25 V that is largely absent in Fig. 6.3b. It is also clear from the figures that the ion currents in RILIS mode and VADIS mode of operation are comparable. Time structure measurements were made at anode voltages corresponding to specific features in the recorded ion current to better understand the phenomena, these are presented in Fig. 6.4.

There is a clear difference in the background of the time structures recorded with 0 and 5 V applied to the anode to those with 2 and 15 V applied. To better facilitate

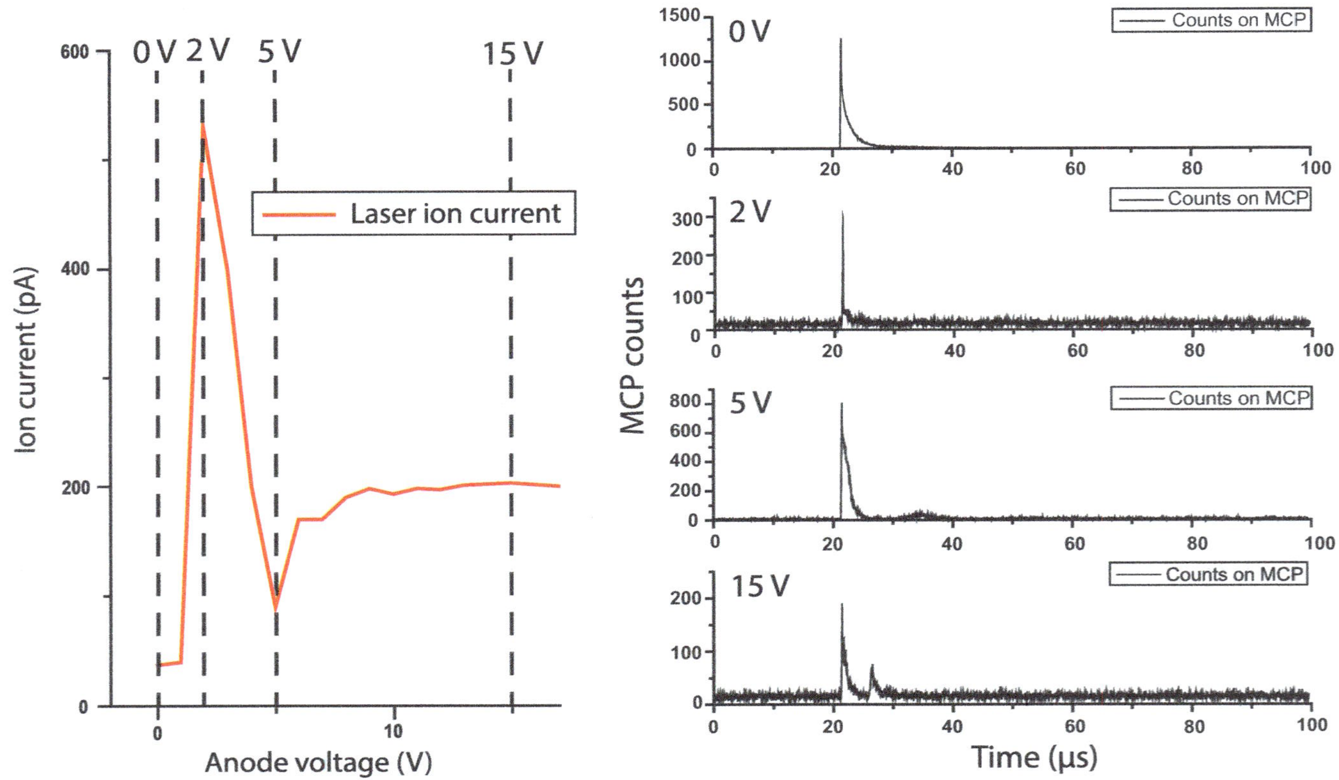

Fig. 6.4 Time structure measurements of the ^{71}Ga laser ionized beam, cathode 340 A. The distinct voltages at which the time structures were obtained are highlighted in the anode voltage scan (detailed view from Fig. 6.3a). Details are given in the text. Figure reproduced from [4] licensed under CC by 4.0

comparison, the contribution of the DC background to the total number of recorded counts was plotted together with the ion currents recorded on the Faraday cup. This is discussed further in [4]. The DC background suggests that prolonged ion survival ($\gg$100 μs) is possible within the anode cavity and that this is the dominant factor for RILIS mode efficiency.

6.5 Further RILIS-Mode Characterization

Reducing the repetition rate of the laser applied for ionization from the standard 10 kHz, facilitates the observation of the laser related ion beam time structure over periods of time >100 μs. Such measurements are presented in Fig. 6.5, where ion beam time structure measurements from a standard HC-RILIS are plotted, with the ionizing laser operating at either 5 or 10 kHz and from RILIS mode VADLIS operation with the ionizing laser operating at 2 kHz. The background observable in the HC-RILIS time structure is predominantly due to surface ionization, whereas in the RILIS mode of VADLIS operation it is nearly 100% laser related. The time structure of RILIS mode VADLIS operation presented in this section was measured with a modified VADIS with an enlarged extraction aperture (from 1.5 to 2.5 mm) to better accommodate the RILIS lasers. Nevertheless, since the change in aperture size did not result in an observed difference in the ion source behaviour (efficiency, parameter optimization), the time structure measurements were deemed relevant.

The time structures were recorded with an MCP, using the experimental set-up and techniques discussed in Sect. 6.4. Two distinct components are visible in the time structures measured in HC-RILIS operation (Fig. 6.5a). The first peak corresponds to the ionization of atoms at the extraction electrode and the mouth of the hot cavity. The second, broader peak, corresponds to atoms ionized along the length of the hot cavity. These ions are accelerated towards the extraction aperture due to the longitudinal potential discussed in Sect. 5.2.1. The HC-RILIS time structure measurements confirm the understanding that the majority of the laser related ions are extracted within the 100 μs duty cycle of the lasers.

Figure 6.5b presents the time structure measurements of the laser ions extracted when operating the VADLIS in RILIS mode, with the ionizing laser operated with a 2 kHz repetition rate. It is clear that the laser ion residency time greatly exceeds 500 μs. The discussion in [4] suggests that the extended residency time compared to that of the hot cavity is the result of a central potential well. Such a potential well had previously been suggested in VADIS mode operation, based on the simulation results of L. Penescu [8] using the charged particle optics software CPO [15]. The ion beam time structure measurements, presented in Figs. 6.4 and 6.5b, support the idea that a similar potential well exists in the anode volume when operating in RILIS mode (anode voltage $\approx$5 V). A new simulation campaign was initiated with the goal of gaining insights into both the RILIS mode of VADLIS operation and FEBIAD sources in general.

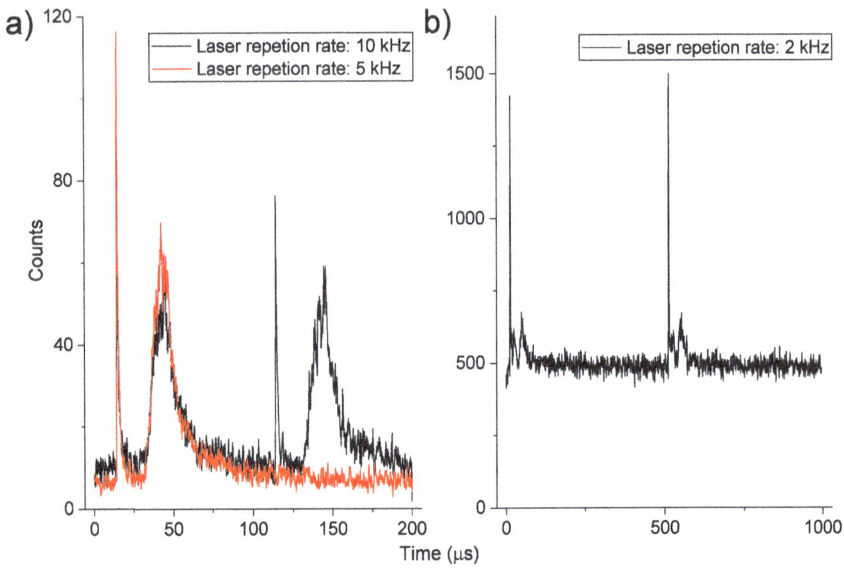

Fig. 6.5 Ion beam time structures from **a** HC-RILIS operation and **b** RILIS mode VADLIS operation

The particle-in-cell (PIC) software VSIM [16] was identified as suitable for this task. A precursor to VSIM (VORPAL) was previously used by L. Penescu to study space charge effects in FEBIAD sources [17]. The potential benefits of VSIM compared to electrostatic simulation software such as CPO, is the capability of including the required physics to model plasmas, electron impact ionization and collisional ionization/neutralization, thus enabling a greater range of the experimental conditions to be taken into account. In PIC simulations a grid or mesh is overlaid on the structures in the simulation. The particles move throughout the simulation region in predefined time steps and at each time step the electromagnetic fields are calculated at the nodes of the grid boundaries and updated. The Courant condition [18, 19] is typically applied, restricting the particles to no more than one grid boundary crossing per time step. As the mass of an electron is 3-5 orders of magnitude lower than that of an atom or ion, its velocity is correspondingly greater. The Courant condition therefore limits the possible duration of the time step, hindering attempts to obtain meaningful results within a reasonable time frame when simulating both ions and electrons based on the available computer resources.

The nature of FEBIAD ion sources, compared to the previously used Modified Nielsen Ion Sources (MNIS) [20], reduced the need for the simultaneous simulation of the electrons with the ions and atoms. The major improvement the FEBIAD offered was the introduction of the anode grid [21], which enables the ion source to operate with a range of internal pressures, down to 10^{-5} mbar and ion currents below 1 μA. The relative insensitivity of the FEBIAD ion sources to variations in the pressure and ion load, implies the possibility to consider an electron dominated regime. In

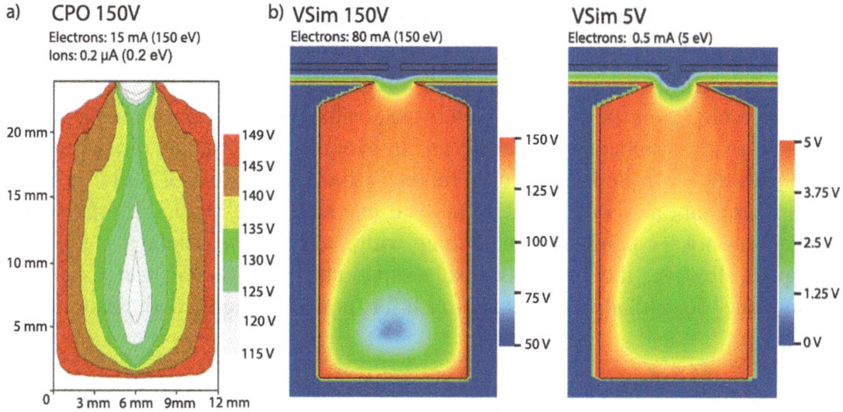

Fig. 6.6 Simulation of the electric potential inside the VADLIS anode cavity **a** in VADIS mode using CPO (from [8]) and **b** VADIS mode and RILIS mode using VSim. Input parameters are discussed in the text

the RILIS mode of operation, the electron current is lower because of the reduction in the voltage applied to the anode. However, the extracted ion current is similarly reduced due to the selective nature of the RILIS mode of operation and does not typically exceed 10 nA.

An initial VSim simulation code was written working in collaboration with J. Smith from TechX. The further development and subsequent application of the code for FEBIAD optimization will form part of the PhD thesis of Y. Martinez. Early results from the VSim simulations are presented in Fig. 6.6b, together with the results of simulations with CPO reported in [8] for comparison (Fig. 6.6a).

For the VSim simulations, electrons were introduced from the rear of the anode cavity at each time step. The anode voltage and electron energies were set as 150 V and 5 V for the VADIS mode and RILIS mode simulations, respectively. The simulations were run until the electron population within the anode cavity reached steady state conditions. Comparing the simulation results for VADIS mode operation between the CPO and VSim simulations, a similar potential well is observable at the rear of the anode cavity, with a similar potential distribution across the anode volume. There were several differences in the input parameters for the VSim simulations compared to the previous CPO simulations. The electron current was increased to 80 mA based on the current measured on the anode voltage supply during operation in VADIS mode and the ion contribution to the electric potential was not considered (based on the capability of the source to operate over a wide range of ion currents). These assumptions require further verification, however, despite this and the simulations being conducted using different software, the results are comparable, suggesting the validity of the approach.

When operating in RILIS mode, the electron current is no longer within the range of measurement using the anode voltage supply. Instead, the electron current resulting from the application of 5 V to the anode was estimated using Child's law for space charge limited electron emission, which states that the electron emission is proportional to $V^{3/2}$. Applying this relationship, the electron current was reduced to 0.5 mA. Comparing the VSim simulations for VADIS mode and RILIS mode operation (Fig. 6.6b), a similar potential well is observable with 5 V applied to the anode. This is in good agreement with the interpretation of the time structure measurements.

The results presented in Fig. 6.6b represent only the results of initial proof of concept simulations. For example, the axial magnetic field created by the source magnet was not included and ion tracking within the electric potential is yet to be attempted. However, the simulation results support the discussion in Sect. 6.4 and [4] and indicate the suitability of this approach for future ion source optimization.

6.6 On-Line Testing and Verification

Following successful off-line testing, an immediate application of the VADLIS developments was the possibility to couple the RILIS with molten metal targets at ISOLDE. The on-line performance of the VADLIS was tested with mercury prior to the experiment reported in Chaps. 8–10 and with radiogenically produced cadmium. Figure 6.7 is a comparison of the VADLIS modes of operation, observed for the two different target units during mercury yield measurements and cadmium ionization tests. The ^{178}Hg rate was measured by α-decay spectroscopy in the CRIS DSS 2.0 setup [22], while the ^{114}Cd signal was measured using a Faraday cup located after the ISOLDE GPS dipole magnet. The on-line tests for mercury and cadmium took place in-between scheduled ISOLDE experiments. The VADIS parameters were therefore maintained at the optimized values for on-line VADIS mode operation.

All RILIS mode operation has thus far been validated by comparison to VADIS mode ion currents, for which the absolute efficiency has not yet been determined. Figure 6.7 demonstrates a comparable efficiency for RILIS mode and VADIS mode operation under on-line conditions. A mercury ionization efficiency of 60% was reported at ISOLDE with the MK6 FEBIAD variant [7]. The MK6 however, differs in anode design and component material compared to the VD3 VADIS variant, currently used to ionize mercury at ISOLDE. An efficiency measurement of the RILIS ionization scheme for mercury in a hot cavity gave a lower limit efficiency of 6% (Sect. 7.4). Figure 6.7 demonstrates the comparable on-line ionization efficiencies of RILIS mode and VADIS mode operation for mercury and cadmium, with the highest ion currents observed in the RILIS + VADIS mode. Figure 6.7 also includes the signal when of directing the 40 W Blaze laser into the VADIS anode (VADIS + Blaze), the increased VADIS mode signal is thought to be due the incident laser beam increasing the temperature of the cathode surface.

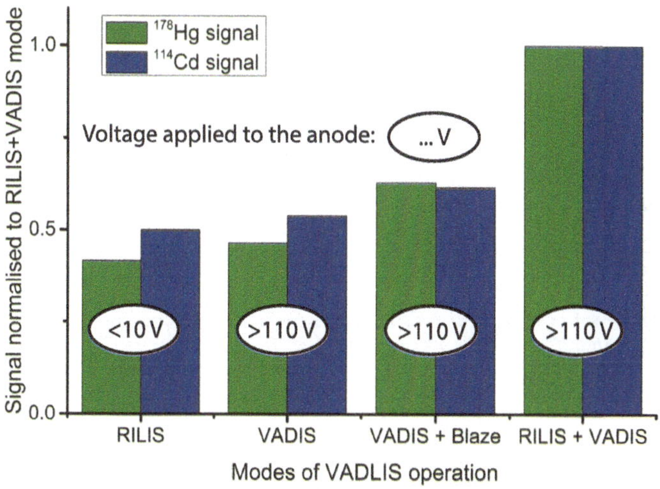

Fig. 6.7 Comparison of the VADLIS modes of operation. The Blaze is a 40 W frequency doubled Nd:YVO_4 laser, typically directed into a cavity for a non-resonant final step transition to the ionization continuum. Figure reproduced from [4] licensed under CC by 4.0

6.7 On-Line Operation

6.7.1 On-Line Application for Mercury

The coupling of the RILIS with molten targets, combined with what is essentially a new type of laser ion source cavity, has already enabled the study of nuclear ground state and isomer properties of mercury isotopes between ^{177}Hg and ^{208}Hg (Chaps. 8–10). The particular advantages of employing the VADLIS-molten lead combination are discussed in Sect. 8.2.

6.7.2 On-Line Application for Magnesium: A Truly Versatile Ion Source

The possibility of switching between RILIS and VADIS modes of operation has a wider significance for ISOLDE operation. This is illustrated by its application on-line at ISOLDE for an ISOLTRAP experiment. Ion beams of magnesium, sodium and neon were required for the study of the Q-values of mirror transitions for fundamental interaction studies [23]. A VD5 "hot plasma" ($\approx$2000 °C transfer line) VADIS was installed, coupled to a SiC target to try to achieve this. However, the contamination on mass $A = 23$ when operating in VADIS mode was too great for the measurement of ^{23}Mg. By switching to the RILIS mode of operation and applying the $\{\lambda_1\,|\lambda_2|\,\lambda_3\} =$

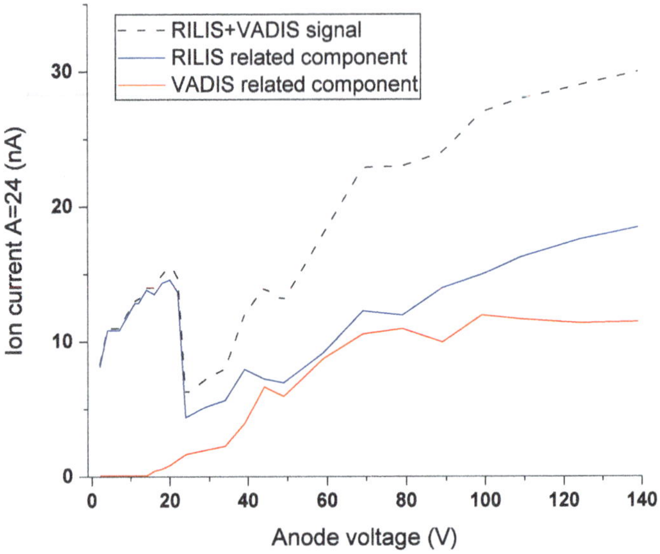

Fig. 6.8 On-line VADLIS anode voltage scan for magnesium

{285 nm|553 nm|532 nm^{NR}} RILIS ionization scheme for magnesium [24], the isobaric contamination was sufficiently reduced and the experiment could take place. Thus it was possible to make effective use of surface ionized, electron impact ionized and laser ionized radioactive ion beams from the same target-ion source unit.

An anode voltage scan was made in a similar manner to those described in Sect. 6.4, with 345 A applied to the cathode (≈1980 °C). However, during this scan, the source magnet was optimized at each measurement point. The results are presented in Fig. 6.8.

There is a smooth increase in the VADIS mode ion current with increasing anode voltage. Considering the laser ionized component, a clear low voltage RILIS mode of operation is once again observable. Under on-line conditions the RILIS mode and VADIS mode were again demonstrated to have to comparable efficiencies, with a doubling of the ion rate possible when operating in the RILIS + VADIS mode.

6.8 Outlook

There are a number of possibilities for future developments ranging from the production of refractory metal ion beams or additionally selective RILIS operation to efficiency measurements, anode cavity optimization and two photon spectroscopy [4]. The comparable RILIS mode and VADIS mode ion signals presented in Sects. 6.4, 6.6 and 6.7.2, indicate a reasonable RILIS mode ionization and extraction efficiency, based on the understanding that the VADIS is a universal ion source [7]. This does not however, imply that the scope for optimization is limited. The time structure measurements indicate the efficiency of RILIS mode operation is linked to an ion survival time of >> 500 μs, compensating for the slow extraction of laser ionized ions. One of the avenues of potential interest is the use of the VADIS mode to break up molecules and then the RILIS mode of operation to selectively ionize the constituent of interest, as discussed in [4]. Ideas for achieving this involve the pulsing of the voltage applied to the anode to provide electrons with sufficient energy for molecular breakup, followed by ionization and extraction in the RILIS mode of operation. Such a method does not appear compatible with the current RILIS mode of operation due to the slow extraction of RILIS related ions. Optimization of the anode cavity to reduce the ion extraction time is therefore crucial for such developments to take place. Additionally, were extraction in the range of 10 μs to be achieved, then it could enable time-gating methods [25] to be applied for enhanced selectivity.

6.9 Conclusion

The coupling of the RILIS with the VADIS has now been demonstrated off-line for gallium and barium and on-line for mercury, cadmium and magnesium, resulting in the VADLIS becoming an established ion source option at ISOLDE. This is the first time a RILIS and a FEBIAD-type ion source have been successfully combined. Three distinct modes of operation are now available: an element selective RILIS mode, the standard VADIS mode and a combined RILIS + VADIS mode for enhanced on-line efficiencies. The immediate applications of these developments include the coupling of the RILIS with molten targets at ISOLDE and the introduction of two new modes of FEBIAD operation: an element selective RILIS mode and a RILIS + VADIS mode for increased efficiency compared to VADIS mode operation alone.

The efficiency appears to depend strongly on the possibility of extracting laser ions from the full anode cavity and is linked to laser-ion extraction times of ≫500 μs from within the anode cavity. Early results from the simulation of the ion source, operating in both RILIS mode and VADIS mode, appear promising and are in agreement with the interpretation of ion beam time structure measurements: that the ion survival and comparatively slow extraction in RILIS mode is the result of a potential well forming at the rear of the anode cavity. The selectivity of the RILIS mode has also been demonstrated [4], potentially widening the range of RILIS applicable elements to those that are typically surface ionized at ISOLDE. The results presented here mark the beginning of further VADLIS development and optimization.

References

1. Blaum K, Geppert C, Kluge H-J, Mukherjee M, Schwarz S, Wendt K (2003) A novel scheme for a highly selective laser ion source. Nucl Instrum Methods Phys Res Sect B: Beam Interact Mater At 204:331–335. https://doi.org/10.1016/S0168-583X(02)01942-0
2. Fink DA, Richter SD, Bastin B, Blaum K, Catherall R, Cocolios TE, Fedorov DV, Fedosseev VN, Flanagan KT, Ghys L, Gottberg A, Imai N, Kron T, Lecesne N, Lynch KM, Marsh BA, Mendonca TM, Pauwels D, Rapisarda E, Ramos JP, Rossel RE, Rothe S, Seliverstov MD, Sjödin M, Stora T, Van Beveren C, Wendt KDA (2013) First application of the laser ion source and trap (LIST) for on-line experiments at ISOLDE. Nucl Instrum Methods Phys Res Sect B: Beam Interact Mater At 317:417–421. https://doi.org/10.1016/j.nimb.2013.06.039
3. Fink DA, Cocolios TE, Andreyev AN, Antalic S, Barzakh AE, Bastin B, Fedorov DV, Fedosseev VN, Flanagan KT, Ghys L, Gottberg A, Huyse M, Imai N, Kron T, Lecesne N, Lynch KM, Marsh BA, Pauwels D, Rapisarda E, Richter SD, Rossel RE, Rothe S, Seliverstov MD, Sjödin AM, Van Beveren C, Van Duppen P, Wendt KDA (2015) In-source laser spectroscopy with the laser ion source and trap: first direct study of the ground-state properties of 217, 219Po. Phys Rev X 5(1):011018. https://doi.org/10.1103/PhysRevX.5.011018
4. Day Goodacre T, Billowes J, Catherall R, Cocolios TE, Crepieux B, Fedorov DV, Fedosseev VN, Gaffney LP, Giles T, Gottberg A, Lynch KM, Marsh BA, Mendonça TM, Ramos JP, Rossel RE, Rothe S, Sels S, Sotty C, Stora T, Van Beveren C, Veinhard M (2016) Blurring the boundaries between ion sources: the application of the RILIS inside a FEBIAD type ion source at ISOLDE. Nucl Instrum Methods Phys Res Sect B: Beam Interact Mater At 376:39–45. https://doi.org/10.1016/j.nimb.2016.03.005
5. Ravn HL, Sundell S, Westgaard L (1975) Target techniques for the ISOLDE on-line isotope separator. Nucl Instrum Methods 123(1):131–144. https://doi.org/10.1016/0029-554X(75)90089-0
6. Bjørnstad T, Hagebø E, Hoff P, Jonsson OC, Kugler E, Ravn HL, Sundell S, Vosicki B, ISOLDE Collaboration (1986) Methods for production of intense beams of unstable nuclei: new developments at ISOLDE. Phys Scr 34(6A):578–590. https://doi.org/10.1088/0031-8949/34/6A/013
7. Sundell S, Ravn H (1992) Ion source with combined cathode and transfer line heating. Nucl Instrum Methods Phys Res Sect B: Beam Interact Mater At 70(1–4):160–164. https://doi.org/10.1016/0168-583X(92)95926-I
8. Penescu L, Catherall R, Lettry J, Stora T (2010) Development of high efficiency versatile arc discharge ion source at CERN ISOLDE. Rev Sci Instrum 81(2):02A906. https://doi.org/10.1063/1.3271245
9. Mishin VI, Fedoseyev VN, Kluge H-J, Letokhov VS, Ravn HL, Scheerer F, Shirakabe Y, Sundell S, Tengblad O (1993) Chemically selective laser ion-source for the CERN-ISOLDE on-line mass separator facility. Nucl Instrum Methods Phys Res Sect B: Beam Interact Mater At 73(4):550–560. https://doi.org/10.1016/0168-583X(93)95839-W
10. Ames F, Brumm T, Jager K, Kluge H-J, Suri BM, Rimke H, Trautmann N, Kirchner R (1990) A high-temperature laser ion source for trace analysis and other applications. Appl Phys B Photophys Laser Chem 51(3):200–206. https://doi.org/10.1007/BF00405739
11. Fedosseev VN, Berg L-E, Fedorov DV, Fink D, Launila OJ, Losito R, Marsh BA, Rossel RE, Rothe S, Seliverstov MD, Sjödin AM, Wendt KDA (2012) Upgrade of the resonance ionization laser ion source at ISOLDE on-line isotope separation facility: new lasers and new ion beams. Rev Sci Instrum 83(2):02A903. https://doi.org/10.1063/1.3662206
12. Schwellnus F, Catherall R, Crepieux B, Fedosseev VN, Marsh BA, Mattolat Ch, Menna M, Österdahl FK, Raeder S, Stora T, Wendt K (2009) Study of low work function materials for hot cavity resonance ionization laser ion sources. Nucl Instrum Methods Phys Res Sect B: Beam Interact Mater At 267(10):1856–1861. https://doi.org/10.1016/j.nimb.2009.02.068
13. Ross KJ, Sonntag B (1995) High temperature metal atom beam sources. Rev Sci Instrum 66(9):4409–4433. https://doi.org/10.1063/1.1145337

14. Kramida A, Ralchenko Yu, Reader J, NIST ASD Team (2015) NIST atomic spectra database (ver. 5.2). http://physics.nist.gov/asd. Accessed 22 July 2016
15. Charged Particle Optics Ltd. CPO (2015). http://simion.com/cpo/
16. TECH-X. VSim (2016). https://www.txcorp.com/vsim
17. Penescu L, Catherall R, Lettry J, Stora T (2008) Numerical simulations of space charge effects and plasma dynamics for FEBIAD ion sources. Nucl Instrum Methods Phys Res Sect B: Beam Interact Mater At 266(19–20):4415–4419. https://doi.org/10.1016/j.nimb.2008.05.071
18. Courant R, Friedrichs K, Lewy H (1928) Über die partiellen Differenzengleichungen der mathematischen Physik. Mathematische Annalen 100(1):32–74. https://doi.org/10.1007/BF01448839
19. Courant R, Friedrichs K, Lewy H (1967) On the partial difference equations of mathematical physics (Republished). IBM J Res Dev 11(2):215–234. https://doi.org/10.1147/rd.112.0215
20. Kirchner R, Roeckl E (1976) A novel ISOL ion source. Nucl Instrum Methods 139:291–296. https://doi.org/10.1016/0029-554X(76)90687-X
21. Kirchner R, Roeckl E (1976) Investigation of gaseous discharge ion sources for isotope separation on-line. Nucl Instrum Methods 133(2):187–204. https://doi.org/10.1016/0029-554X(76)90607-8
22. Lynch KM, Cocolios TE, Althubiti N, Farooq-Smith GJ, Smith AJ, A simple chamber for decay spectroscopy at CRIS-ISOLDE. Unpublished results
23. Breitenfeldt M, Atanasov D, Blaum K, Eronen T, Finlay P, Herfurth F, Kowalska M, Kreim S, Litvinov Yu, Lunney D, Manea V, Neidherr D, Porobic T, Rosenbusch M, Schweikhard L, Severijns N, Wienholtz F, Wolf RN, Zuber K (2013) Q-values of mirror transitions for fundamental interaction studies. Technical report, CERN-INTC, Geneva. http://cds.cern.ch/record/1551494
24. Köster U, Fedoseyev VN, Andreyev AN, Bergmann UC, Catherall R, Cederkäll J, Dietrich M, De Witte H, Fedorov DV, Fraile L, Franchoo S, Fynbo H, Georg U, Giles T, Gorska M, Hannawald M, Huyse M, Joinet A, Jonsson OC, Kratz KL, Kruglov K, Lau Ch, Lettry J, Mishin VI, Oinonen M, Partes K, Peräjärvi K, Pfeiffer B, Ravn HL, Seliverstov MD, Thirolf P, Van de Vel K, Van Duppen P, Van Roosbroeck J, Weissman L (2003) On-line yields obtained with the ISOLDE RILIS. Nucl Instrum Methods Phys Res Sect B: Beam Interact Mater At 204:347–352. https://doi.org/10.1016/S0168-583X(02)01956-0
25. Rothe S, Catherall R, Crepieux B, Day Goodacre T, Fedosseev VN, Giles T, Marsh BA, Ramos JP, Rossel RE (2016) Advances in surface ion suppression from RILIS: towards the time-of-flight laser ion source (ToF-LIS). Nucl Instrum Methods Phys Res Sect B: Beam Interact Mater At 376:86–90. https://doi.org/10.1016/j.nimb.2016.02.060

Chapter 7
Ionization Scheme Development for the ISOLDE RILIS

Abstract Ionization scheme development enables new or more efficient ion beam production by the ISOLDE RILIS. The selective nature of the ionization process requires a scheme to be developed for each element of interest. This chapter presents the methodology of ionization scheme development and new ionization schemes for tellurium, germanium, chromium, mercury and radium.

7.1 Ionization Scheme Development

The developments presented in this chapter took place at ISOLDE (Sect. 5.1) using the RILIS system presented in Sect. 5.2.2. The procedure for ionization scheme development for the ISOLDE RILIS was reviewed in detail by V.N. Fedosseev et al. [1]. The process typically involves a sequence of distinct investigations:

- A review the known spectral information and the determination of the atomic distribution between the ground state and thermally populated energy levels in the hot cavity environment (Sect. 7.1.1).
- Laser spectroscopy, applying the RILIS laser system to verify the existing spectral information, investigating new lines and the optimal ionization scheme (Sect. 7.1.2).
- Saturation measurements of candidate transitions using the RILIS lasers to determine the laser power required to achieve saturation (Sect. 7.1.3).
- Ionization efficiency measurements (Sect. 7.1.4), this value typically determines whether future spectroscopic investigations will be necessary.

7.1.1 Review and Application of Existing Spectroscopic Information

The process begins with an assessment of the available literature data on atomic lines and energy levels. The primary sources of information are the Kurucz and NIST atomic line databases [2, 3] and the extensive work of E.B. Saloman [4–10].

T. Day Goodacre, *Applied Laser Spectroscopy for Nuclear Physics*, Springer Theses,
https://doi.org/10.1007/978-3-030-73889-1_7

There are a number of factors that determine the applicability of an atomic energy level or transition for use within an ionization scheme.

In the hot cavity environment the atomic population may be divided among a number of thermally populated energy levels in addition to the atomic ground state. The percentage of atoms P_a populating a state a can be calculated by applying the Boltzmann equation in the form given in Eqs. (7.1)–(7.3):

$$\theta = g_a exp(-E_a/kT) \tag{7.1}$$

$$\Theta = \sum_i g_i exp(-E_i/kT) \tag{7.2}$$

$$P_a = (\theta/\Theta) \times 100 \tag{7.3}$$

where g is the degeneracy of the atomic state, E is the energy of the atomic state, k is the Boltzmann constant, T is the temperature of the hot cavity in kelvin (K) and the subscript i represents the group of atomic states with a significant thermal population at temperature T.

Where possible, an excitation path via states with progressively increasing total angular momentum (J) is preferable. This is because the theoretical maximum efficiency of an excitation is greater if the statistical weight (2J+1) of the upper level of the atomic excitation is greater than that of the lower level [11].

The saturation of all of the transitions applied in an ionization scheme both maximises the achievable efficiency and corresponds to a reduced sensitivity to any potential deterioration in the RILIS laser power over the course of the experiment. The possibility to saturate a candidate transition with the available laser power can be estimated by considering literature data on transition strengths. Candidate transitions are compared to transitions used in existing RILIS ionization schemes, by employing Eqs. (4.1)–(4.3).

7.1.2 Atomic Spectroscopy

A spectroscopic investigation is undertaken with the starting point based on the available literature data. A mass marker containing a sample of the element of interest equivalent to ≈10 000 nAh (assuming a 100% conversion to singly charged ions) is used to ensure a sufficient supply of atoms. The known transitions are verified by directing the lasers into the hot cavity while heating the mass marker and detecting the ions on a Faraday cup inserted in the focal plane of the dipole mass separator.

Following the verification of the known transitions, a tunable final step laser is introduced to search for autoionizing states or to investigate Rydberg levels. While the final step laser is scanned, the Faraday cup measurements are combined with the laser wavelengths measured with a HighFinesse/Ångstrom WS7 wavelength meter. The compilation and recording of experimental parameters is accomplished with a LabVIEW based data recording system [12, 13]. The purpose of the spectroscopic investigations is the identification of an optimal RILIS ionization scheme for ion beam production at ISOLDE, RILIS operating parameters (laser linewidths and power levels) are therefore optimized for ionization efficiency rather than spectral resolution.

7.1.3 Saturation Measurements

The saturation requirement of a resonant transition is determined by varying the laser power delivered to the ion source for the transition and monitoring the ion current, while maintaining all other experimental parameters (laser beam positions and wavelengths, ion source and mass marker heating and the laser power delivered for the other applied transitions) constant. Equation (7.4) is often used to fit a curve to the saturation data. This function relates the signal to the laser power delivered for each resonant transition by determining a saturation parameter $P_{1/2Sat}$ [14, 15]:

$$I = B + M(P/P_{1/2Sat})/(P/(P_{1/2Sat} + 1)) \tag{7.4}$$

Where I is the ion current, B is the background ion current when the transition is blocked, M is the maximum ion current achievable with a fully saturated transition, P is the laser power delivered to the hot cavity, $P_{1/2Sat}$ is the laser power required to achieve 50% of M. The equation assumes that the dominant contribution to the ionization process is from resonant excitation.

7.1.4 Efficiency Measurements

The absolute efficiency of the ionization scheme corresponding to the highest ion signal is measured in order to determine the suitability of the scheme for use at ISOLDE. The time available for tests and experiments using the ISOLDE front-ends, mass separators and beam lines is limited due to the annual demand for physics, development, maintenance or upgrades. This typically results in just one efficiency measurement per element being possible. In the standard (target-transfer line-hot cavity) configuration used for on-line ISOLDE experiments, there is an additional laser-atom overlap region in the transfer line that increases the efficiency of the ionization process [16]. To conserve resources, efficiency measurements typically take place with a simplified mass marker-hot cavity configuration. The measured efficiencies could therefore be considered as a lower limit of the ionization efficiency.

To determine the efficiency of an ionization scheme, a calibrated mass marker of typically between 1000 and 4000 nAh is evaporated, while ionization scheme is applied and the ion current is measured with a Faraday cup inserted downstream of the dipole mass separator. A Faraday cup measurement is recorded periodically throughout the evaporation of the sample. These measurements are used to calculate the average recorded signal over the evaporation time and thus the efficiency of the ionization scheme. The final efficiency value is a combination of the ionization, extraction and ion beam transport efficiency. No attempt is made to correct for these factors as they are intrinsic to the system, thus a corrected value would not provide a realistic determination of the achievable efficiency of on-line ion source operation.

7.2 Chromium

The work presented in this section was published in Spectrochimica Acta Part B: Atomic Spectroscopy, in a paper titled "*The identification of autoionizing states of atomic chromium for the resonance ionization laser ion source of the ISOLDE radioactive ion beam facility*" [17]. The paper is licensed under CC by 4.0.

7.2.1 Motivation

The development of a RILIS ionization scheme for chromium was motivated by a request from the ISOLTRAP collaboration for RILIS ionized neutron rich chromium isotopes for mass measurements at ISOLDE [18] and additional interest from the ISOLDE solid state physics group to investigate the use of ^{48}Cr as a magnetic probe for Perturbed Angular Correlation (PAC) spectroscopy (private communication Karl Johnston, ISOLDE Physics coordinator). Chromium had previously been ionized at ISOLDE using "hot plasma" variants of FEBIAD ion sources coupled with a ZrO_2 or an Y_2O_3 target [19]. The "universal" ionization efficiency of the hot plasma FEBIAD limits the achievable purity of the produced radioactive ion beams, which can become a particularly significant problem if a UC_x target is used [20]. Furthermore, the VADIS anode cavity is only heated radiatively and by electron bombardment, rather than by direct ohmic heating as used for the hot cavity. The effect of wall sticking times can result in isotope (half-life) specific ion source efficiencies, particularly for less-volatile elements such as chromium. For the more exotic chromium isotopes, the achievable VADIS ionization efficiency may therefore be hindered more by wall sticking times than the achievable efficiency of RILIS ionization in a hot cavity.

HC-RILIS operation therefore offers:

- Increased purity,
- Signal identification by blocking or detuning one of the resonant transitions,
- A potentially reduced release time.

7.2.2 *Ionization of Chromium Atoms in a Hot Cavity*

By inputting information from the NIST database [3] into Eqs. (5.1) and (5.2) the effective first ionization potential of chromium can be calculated to be 6.8 eV at 2400 K. Consequently, with sufficient heating of a mass marker containing stable chromium, a surface ion signal measurable on an ISOLDE Faraday cup (>100 fA) was observable from a tungsten hot cavity ion source heated to 2400 K.

Using previously obtained data for scandium (RILIS and surface ionization efficiencies) [1] and Eqs. (5.3) and (5.4), it was possible to estimate ξ to be $\approx$0.1 in a tungsten hot cavity heated to 2400 K. The assumptions for the applicability of this value to chromium are discussed in [17]. By inputting $\xi = 0.1$ into Eqs. (5.3) and (5.4), a surface ionization efficiency of chromium in a tungsten hot cavity was calculated to be $\approx$0.01%.

7.2.3 *Existing Spectroscopic Information*

Based on Eqs. (7.1)–(7.3) and data from the NIST [3] and the Kurucz databases [2], the $3d^5(^6S)4s\ a^7S_3$ atomic ground state of chromium is 97% populated at 2400 K. Thus only ionization schemes originating from the atomic ground state were considered. Extensive atomic line data is available for chromium, including numerous autoionizing states [2, 3, 10]. Suitable first and second step transitions were identified that excited to an energy of 42 261.226 cm^{-1}, $\approx$12 314 cm^{-1} ($\approx$812 nm) below the ionization threshold [3]. The details of these transitions are presented in Table 7.1.

7.2.4 *Laser Spectroscopy*

Light for the first and second step transitions was produced by frequency doubling the output of a Ti:Sa laser and a dye laser operated with an ethanol solution of Pyridine 1. The transitions listed in Table 7.1 were initially applied as part of a $\{\lambda_1 | \lambda_2 | \lambda_3\} = \{358\text{ nm}|698\text{ nm}|532\text{ nm}^{NR}\}$ ionization scheme with a non-resonant (^{NR}) final step, to verify the suitability of the transitions found in literature. This ionization scheme corresponded to a laser to surface ion ratio of 1000:1.

Table 7.1 The first and second resonant transitions used in the chromium scheme development originating from the $3d^5(^6S)4s\ a^7S_3$ ground state. Spectroscopic information is taken from the NIST database [3]

Transition (cm^{-1})	Upper state config.	Air wavelength (nm)
0 – 27 935.241	$3d^4(^5D)4s4p^3P^o\ y^7P^o_4$	357.869
27 935.241 – 42 261.226	$3d^5(^6S)4d\ e^7D_5$	697.840

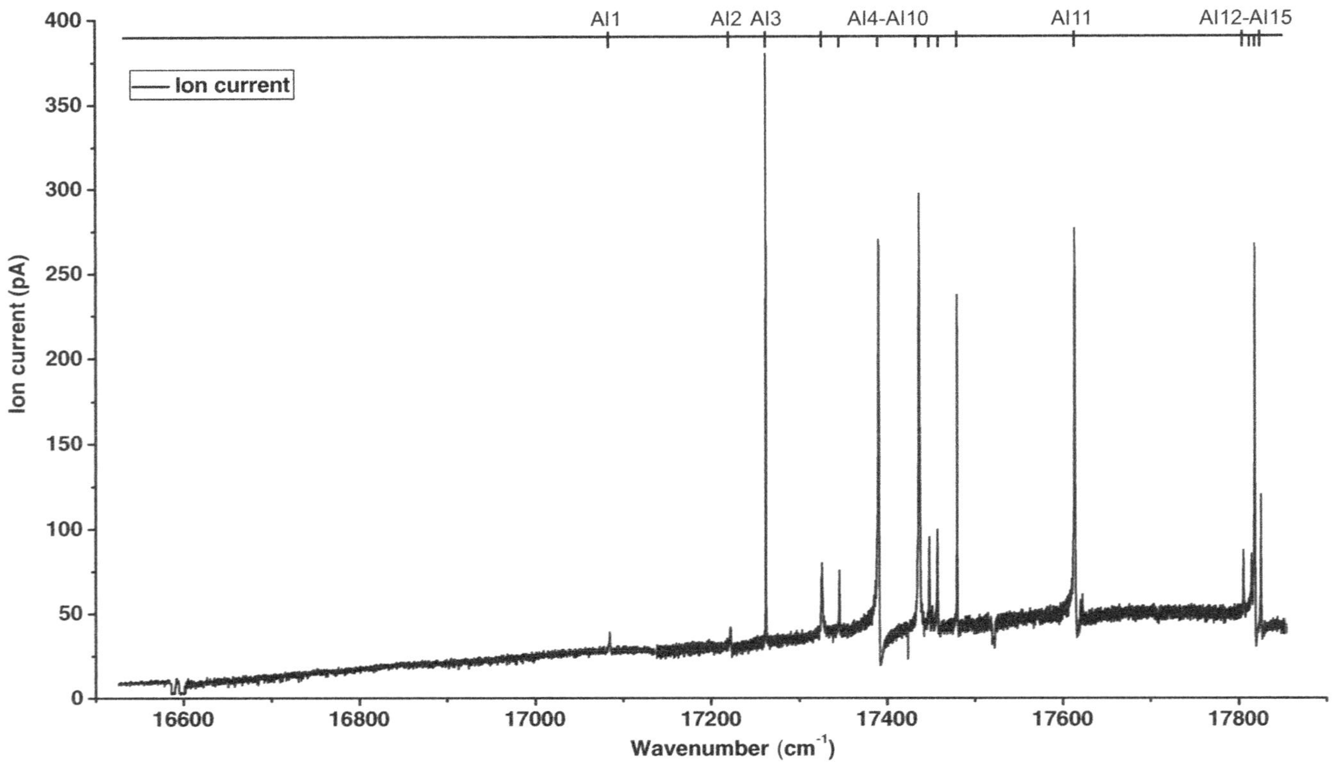

Fig. 7.1 Autoionizing resonances of chromium. The combined spectra from laser scans between 16 526 and 17 854 cm^{-1}, the observed autoionizing resonances were accessed via a single photon transition from the $3d^5(^6S)4d\ e^7D_5$ state at 42 261.226 cm^{-1}. Figure adapted from [17], licensed under CC by 4.0

Table 7.2 Autoionizing states of chromium accessed via a single photon transition from the $3d^5(^6S)4d\ e^7D_5$ level at 42 261.226 cm^{-1}. The optimal ionization scheme is highlighted with a bold font. Air wavelengths calculated using the equation of [21]. No attempt was made to deconvolve the ~12 GHz linewidth of the scanning laser from the full width half maximum (FWHM). Data reproduced from [17]

Identifier	Energy (cm^{-1})	Transition energy (cm^{-1})	Air wavelength (nm)	Relative signal	FWHM (cm^{-1})
AI1	59 345.91 (6)	17 084.68 (6)	585.157	4 (1)	1.11 (2)
AI2	59 483.52 (6)	17 222.29 (6)	580.482	10 (2)	0.92 (8)
AI3	**59 523.39 (3)**	**17 262.16 (3)**	**579.141**	**100**	**1.02 (1)**
AI4	59 587.00 (8)	17 325.77 (8)	577.015	28 (2)	2.1 (2)
AI5	59 606.7 (3)	17 345.5 (3)	576.357	12 (2)	1.5 (3)
AI6	59 652.25 (4)	17 391.02 (4)	574.850	80 (8)	1.48 (3)
AI7	59 697.74 (6)	17 436.51 (6)	573.350	84 (8)	1.85 (3)
AI8	59 709.5 (1)	17 448.3 (1)	572.963	12 (3)	1.1 (2)
AI9	59 718.6 (2)	17 457.4 (2)	572.665	0.20 (1)	1.1 (1)
AI10	59 740.93 (3)	17 479.70 (3)	571.933	40 (1)	0.67 (2)
AI11	59 874.8 (2)	17 613.6 (2)	567.587	65 (6)	1.0 (1)
AI12	60 066.4 (5)	17 805.2 (5)	561.478	12 (2)	0.86 (9)
AI13	60 076.0 (7)	17 814.8 (7)	561.175	8 (2)	0.64 (8)
AI14	60 079.8 (2)	17 818.6 (2)	561.055	63 (6)	0.71 (8)
AI15	60 086.8 (3)	17 825.6 (3)	560.835	25 (3)	0.61 (8)

The 532 nm laser was then replaced with a second dye laser, operated with an ethanol solution of either Rhodamine 6G, Pyrromethene 597 or DCM dye to excite from the $3d^5(^6S)4d\ e^7D_5$ state and scan above the ionization potential between 16 526 and 17 854 cm^{-1} to search for autoionizing resonances. The typical average power ranged between 0.5–10 W across the lasing range of the dyes. The ^{52}Cr ion current was measured while recording the scanning laser wavelength using the RILIS data acquisition system [12]. 15 transitions to new autoionizing states were observed in the scans between 16 526 and 17 854 cm^{-1} (560–605 nm), the combined spectra are presented in Fig. 7.1. The autoionizing resonances were fitted using either Voigt or Breit–Wigner–Fano profiles. The details of the observed resonances are presented in Table 7.2 and are discussed in [17]. The optimal {358 nm|698 nm|579 nm^{AI}} ionization scheme, ionizing via an autoionizing state (^{AI}) is presented in Fig. 7.2, together with laser scans of the three resonances.

The laser power required to saturate the {358 nm|698 nm|579 nm^{AI}} ionization scheme, applying an autoionizing final step, was investigated in the manner described in Sect. 7.1.3. The results of the saturation measurements are also presented in Fig. 7.2. The uncertainty of the values of $P_{1/2sat}$ is discussed in [17]. Despite not being fully saturated, the ion signal achieved when applying the {358 nm|698 nm|579 nm^{AI}} ionization scheme, was a factor of two greater than the ion signal observed when using the {358 nm|698 nm|532 nm^{NR}} ionization scheme with an estimated 24 W of 532 nm laser light transmitted to the ion source.

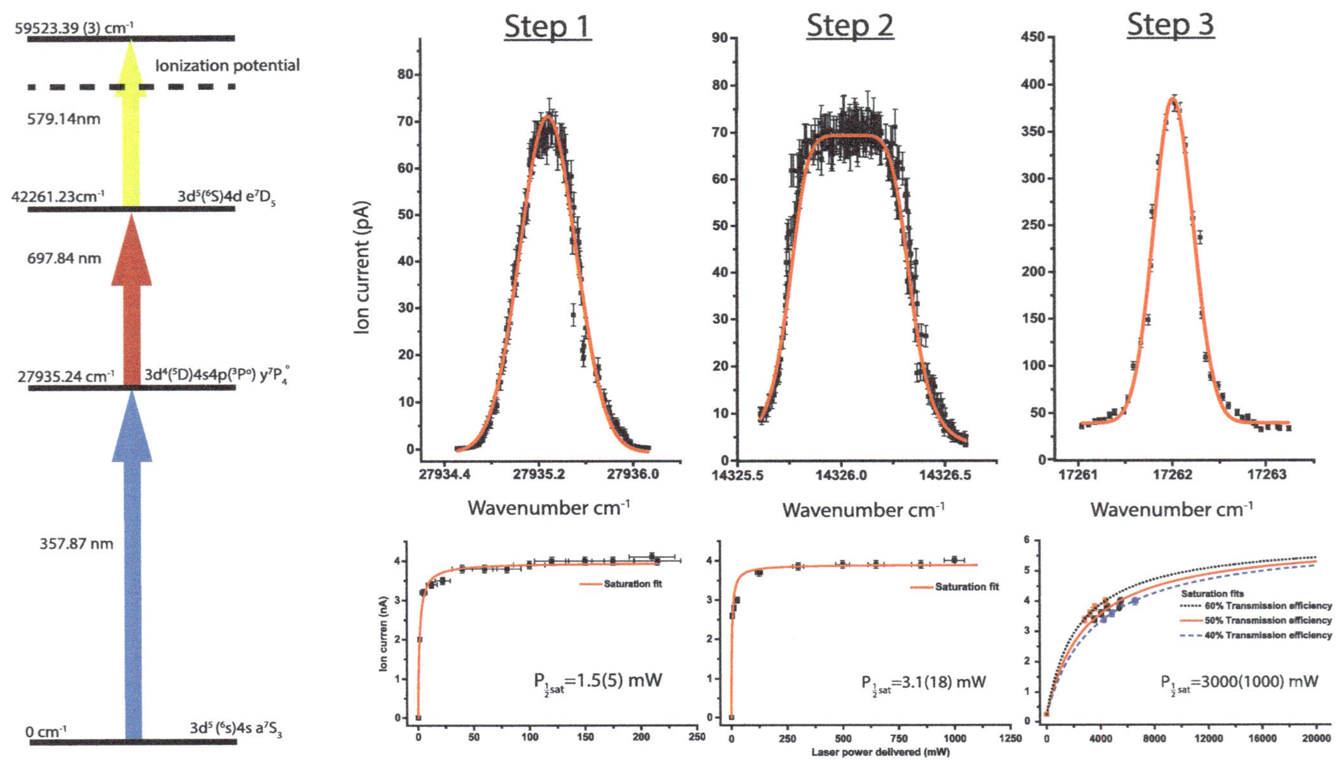

Fig. 7.2 The optimal RILIS ionization scheme for chromium, saturation measurements and scans of the three resonant transitions. The error bars of the laser scan data points are estimated based on the precision of the Faraday cup and the method of data acquisition. The error bars on the individual measurement points of the saturation measurements are related to the precision of the measurement devices and the method of recording, the uncertainty in the value of $P_{1/2sat}$ is discussed further in [17]. Figure adapted from [17], licensed under CC by 4.0

7.2.5 *Efficiency*

There were two unsuccessful attempts to measure the efficiency of the {358 nm| 698 nm|579 nm^{AI}} ionization scheme, using the methodology presented in Sect. 7.1.4. Unfortunately it was not possible to extract meaningful data from these attempts due to excessive amounts of additional chromium present in the target ion source assembly. In an attempt to benchmark the efficiency of the ionization scheme, the RILIS enhancement of the surface ionized chromium ion current on mass A=52 was considered. Using the ionization scheme depicted in Fig. 7.2, a laser to surface ionization ratio of 2200:1 was recorded with the hot cavity heated to 2400 K. Considering the estimated surface ionization efficiency for Cr of ~0.01%, from Eqs. (5.1)–(5.4), a factor of 2200 enhancement of the surface ion current would indicate a RILIS ionization scheme with an efficiency in the region of 20%. This is further discussed in [17].

7.2.6 *On-Line Application at ISOLDE*

The new RILIS ionization scheme has been applied for ISOLTRAP experiments in 2015 and 2016, which measured the mass of neutron rich chromium isotopes as far as ^{63}Cr [22]. Chromium extraction from ISOLDE UC_x targets has been verified, demonstrating the possibility of extracting chromium isotopes with half lives of the order of 100 ms.

7.2.7 *Conclusion*

15 new autoionizing states of chromium have been observed during off-line testing at the ISOLDE RILIS. From this work, an optimal three-step three-resonance {358 nm|698 nm|579 nm^{AI}} RILIS ionization scheme was identified, with an estimated efficiency of 20%. The ionization scheme was applied on-line for the production of neutron rich chromium isotopes as exotic as ^{63}Cr for mass measurements by the ISOLTRAP collaboration. Future possible applications of the ionization scheme for the production of RIB at ISOLDE include collinear spectroscopy and PAC experiments.

7.3 Germanium

The work presented in this section was published in Nuclear Instruments and Methods in Physics Research Section A: Accelerators, Spectrometers, Detectors and Associated Equipment, in a paper titled "*Laser resonance ionization scheme development for tellurium and germanium at the dual Ti:Sa-Dye ISOLDE RILIS*" [23]. The paper is licensed under CC by 4.0.

7.3.1 Motivation

RILIS ionized germanium ion beams were requested for the study of the β^+/EC decay of ^{64}Ge and ^{66}Ge by total absorption spectroscopy [24]. A three-step, Ti:Sa only, laser ionization scheme with a final step transition to an autoionizing state was previously tested and developed at ORNL (Oak Ridge National Laboratory), an efficiency of 3.3% was reported for stable germanium [25].

7.3.2 Existing Spectroscopic Information

There is a significant amount of existing spectroscopic information available for germanium in both the NIST [3] and the Kurucz databases [2]. Based on Eqs. (7.1)–(7.3), the majority of the thermal population at 2300 K is distributed between three atomic states, with 40% lying in the $4s^24p^2\ ^3P_1$ state at 557.134 cm^{-1} and 39% lying in the $4s^24p^2\ ^3P_2$ state at 1409.961 cm^{-1}. Consequently, ionization schemes starting from either $4s^24p^2\ ^3P_1$ or $4s^24p^2\ ^3P_2$ were considered.

7.3.3 Laser Spectroscopy

The 275 nm transition ($4s^24p^2\ ^3P_2 \longrightarrow 4s^24p5s\ ^3P_1^\circ$) was selected as a first step because of an expectation of saturation and a wavelength that could be produced by frequency doubling the light from a dye laser operated with an ethanol solution of

Table 7.3 The resonant transitions tested during the germanium scheme development. Spectroscopic information is taken from [3]. †Ion current measured with ≈24 W of 532 nm light delivered to the ion source for the third, non-resonant, ionization step

Transition (cm^{-1})	Upper state config.	Air wavelength (nm)	Relative ion current
1 409.961 – 37 702.305	$4s^24p5s\ ^3P_1^\circ$	275.459	–
37 702.305 – 54 935.848	$4s^24p6p\ ^3D_1$	580.103	$0.10^\dagger$
37 702.305 – 55 235.834	$4s^24p6p\ ^1P_1$	570.178	$0.27^\dagger$
37 702.305 – 55 266.090	$4s^24p6p\ ^3D_2$	569.196	$1^\dagger$
37 702.305 – 55 503.203	$4s^24p6p\ ^3P_0$	561.614	$0.26^\dagger$

Rhodamine 6G and pumped at 532 nm. From the $4s^24p5s\ ^3P^\circ_1$ state, four second-step transitions, also documented in literature, were observed and compared as part of {275 nm|562 nm, 569 nm, 570 nm, 580 nm|532 nm^{NR}} ionization schemes. The results are summarized and compared in Table 7.3 and the optimal RILIS ionization scheme and laser scans of the first two steps are presented in Fig. 7.3.

The measurements indicated that the first (275 nm) and the optimal second-step transition (569 nm) were both strongly saturated with 580 mW and 14 W respectively on the laser table, estimated to correspond to 300 mW and 5.6 W delivered to the ion source. The spectroscopy is further discussed in [23].

7.3.4 *Efficiency*

An efficiency measurement was performed for the {275 nm|569 nm|532 nm^{NR}} ionization scheme depicted in Fig. 7.3 using a calibrated mass marker of 3 700 nA of germanium (equal to 286 nAh of ^{76}Ge) and the methodology presented in Sect. 7.1.4. The ion current of ^{76}Ge was monitored to ensure there was no influence from the tails of neighbouring stable germanium isotopes. The results are presented in Fig. 7.4.

The ion current was recorded every 10 s with an average signal of 1.41 nA recorded over a period of 3.09 h. Considering the natural isotopic abundance for ^{76}Ge, this corresponds to an efficiency of 1.5%. Due to time constraints, the measurement was terminated before the sample had been fully evaporated however, there was sufficient experimental data to enable the extrapolation of the ion current. A conservative value for the efficiency was determined by constraining the extrapolation to reach zero. An unconstrained extrapolation converged on a 370 pA ion current, equivalent to 0.13% in efficiency per hour and suggesting a significant quantity of un-evaporated germanium remaining in the system. The sum of the measured ion current and the constrained extrapolation (depicted in Fig. 7.4) gives a lower limit estimate for the laser ionization scheme efficiency of 2%.

The addition of a second first step to target atoms in the $4s^24p^2\ ^3P_1$ thermally populated state would have the potential to double the ionization efficiency. This strategy is regularly employed at the ISOLDE RILIS for the ionization of gallium and indium.

7.3.5 *Conclusion*

A new three-step, two-resonance ({275 nm|569 nm|532 nm^{NR}}) RILIS ionization scheme for germanium was identified and tested using the ISOLDE RILIS. A laser ionization efficiency of >2% was measured. This efficiency can be increased by the application of an additional first-step to access another thermally populated low-lying

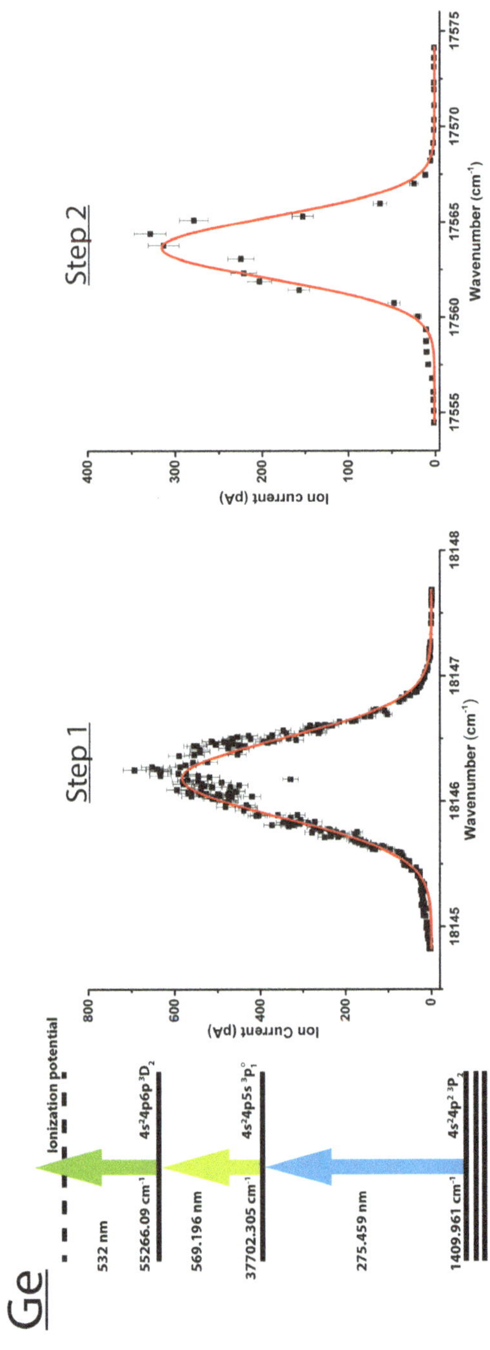

Fig. 7.3 The most efficient of the tested germanium ionization schemes and scans of the resonant transitions. Figure adapted from [23], licensed under CC by 4.0

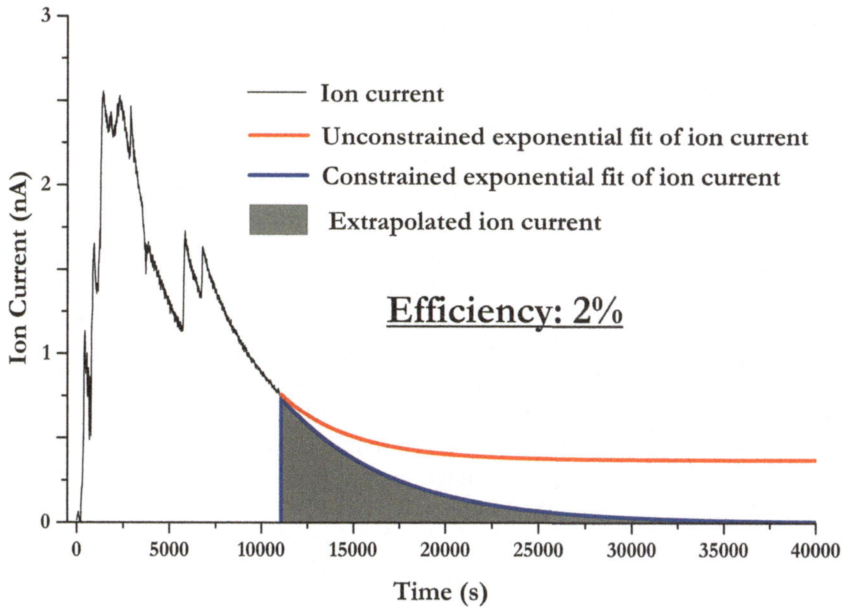

Fig. 7.4 Laser ionized ^{76}Ge ion current during the efficiency measurement. Figure reproduced from [23], licensed under CC by 4.0

atomic energy level, possibly increasing the efficiency of the ionization scheme to >4%.

7.4 Mercury

The work presented in this section was published in Hyperfine Interactions, in a paper titled "*RILIS-ionized mercury and tellurium beams at ISOLDE CERN*" [26]. The paper is licensed under CC by 4.0.

7.4.1 Motivation

A more efficient ionization scheme was required to facilitate an in-source resonance ionization spectroscopy experiment of mercury isotopes, the results of which are presented in Chaps. 8–10.

7.4.2 Existing Spectroscopic Information

There is an extensive range of information available on the atomic structure of mercury in both the NIST [3] and the Kurucz databases [2]. Inputting this information into Eqs. (7.1)–(7.3), 99.85% of the atomic population of mercury resides in the ground state at 2300 K. Thus all of the ionization schemes considered originated from the atomic ground state. A pre-existing {254 nm|313 nm|626 nm} ionization scheme for mercury, was based on a publication by A.A. Podshivalov et al. [27]. The ionization scheme uses the residual 626 nm fundamental light from the 2ω conversion process (to produce the 313 nm second step) as the final ionizing transition. It was suggested that the 626 nm light accessed a favourable part of the ionization continuum due to the vicinity of autoionizing states. The ionization scheme was determined to have an efficiency of 0.1% when tested at the ISOLDE RILIS [28]. By comparison, the reported efficiency of the FEBIAD type ion source for mercury is 60% [29]. Prior to this work, there was therefore limited interest in the application of the RILIS ionization scheme on-line at ISOLDE.

7.4.3 Laser Spectroscopy

A combination of Ti:Sa and dye lasers were used for the investigation of mercury ionization schemes. The first step 254 nm laser light was produced by 3ω frequency conversion of light from a Ti:Sa laser and the second step 313 nm light by 2ω frequency conversion of light produced by a Sirah Credo Dye laser, operated with an ethanol solution of DCM dye. The region around 87 293 cm^{-1} (the excitation energy reached by the A.A. Podshivalov ionization scheme) was investigated, accessed via {254 nm|312.57 nm, 313.16 nm, 313.18 nm|DCM dye scan} ionization schemes. No favourable areas of the continuum were observed, thus a wider ranging spectroscopic investigation using ethanol solutions of either Rhodamine 6G, DCM or Phenoxazone 9 dyes was launched. Rydberg levels were also investigated, exciting from the $5d^{10}6s6d$ $^1D_2$ state. Finally, the effect of the addition of a non-resonant transition at 532 nm was tested, again exciting from each of the second steps, this is presented in Fig. 7.5 together with the resonances corresponding to the first step, three second steps and the Rydberg level that corresponded to the largest ion signal. The spectroscopic information related to the investigated ionization scheme is presented in Table 7.4 and further discussed in [26].

No autoionizing resonances were identified and the 88 759.6 cm^{-1} autoionizing state reported in literature [3] was not observed. There was no barrier to accessing these states based on the reported odd parity and J=1. The saturation of the second steps were verified. Saturation measurements of the first step were not fully conclusive, though saturation effects were observed with an estimated 15 mW of laser power delivered to the ion source.

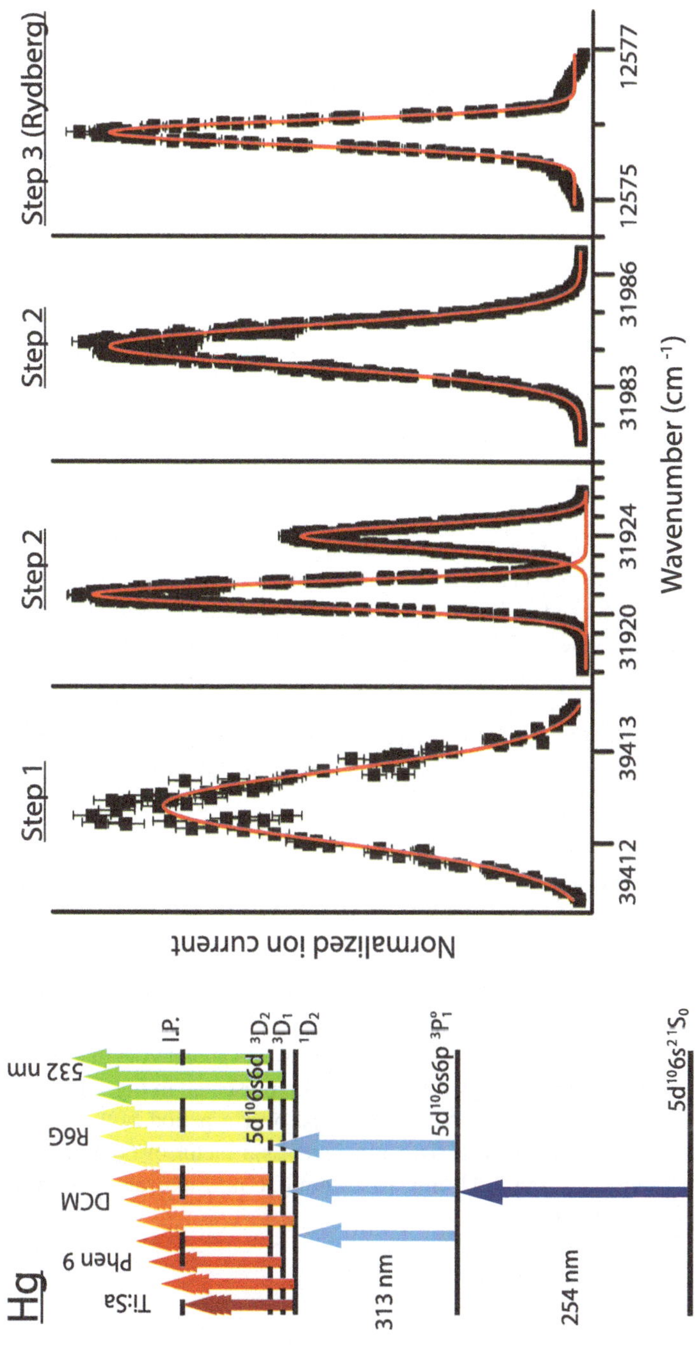

Fig. 7.5 Schematic overview of the ionization scheme development for mercury and scans of the first step, three second steps and the optimal Rydberg resonance. Figure reproduced from [26], licensed under CC by 4.0

Table 7.4 The ionization schemes originated from the $5d^{10}6s^2\ {}^1S_0$ atomic ground state. Spectroscopic information for known transitions is taken from [3]. The given laser power is an estimation of the power delivered to the ion source. *Transitions applied for the efficiency measurement. †Air wavelength calculated using the equation of [21]. Data reproduced from [26]

Transition (cm^{-1})	Upper state config.	Air wavelength (nm)	Laser power (mW)
0 – 39 412.237*	$5d^{10}6s6p\ {}^3P_1^{\circ}$	253.652	80
39 412.237 – 71 333.053*	$5d^{10}6s6d\ {}^1D_2$	313.184	120
39 412.237 – 71 336.005	$5d^{10}6s6d\ {}^3D_1$	313.155	120
39 412.237 – 71 396.073	$5d^{10}6s6d\ {}^3D_2$	312.567	120
71 333.053 – 83 908.95 (4)	Rydberg	794.955† (25)	1 200
71 333.053 – ~90 121*	–	532	24 000

Ionization with the non-resonant final step exciting from the $5p^3(^4S^{\circ})7p\ {}^1D_2$ atomic state was determined to be four times more efficient than via the Rydberg level. This was achieved with an estimated 24 W of 532 nm laser light transmitted to the ion source from the Nd:YVO_4 Lumera Blaze laser [30], compared to an estimated 1.2 W of 795 nm laser light from a Ti:Sa laser. This discrepancy in the signal to power ratio can be understood as resulting from a higher cross-section of excitation to the Rydberg level. The {254 nm|313.18 nm|532 nm^{NR}} ionization scheme was therefore determined to be the optimal one for application at the ISOLDE RILIS.

7.4.4 *Efficiency*

The efficiency of the {254 nm|313.18 nm|532 nm^{NR}} ionization scheme was measured following the methodology of Sect. 7.1.4. The laser parameters used for the efficiency measurement are indicated with an * in Table 7.4. A calibrated sample of 1 336 nAh of mercury (≈400 nAh of ^{202}Hg) was evaporated with a Faraday cup measurement recorded every 10 s. The results are presented in Fig. 7.6.

An average signal of 13.2 nA was recorded over a period of 1.88 h, this corresponds to an efficiency of 6%. This efficiency is a factor of 60 greater than the measured efficiency of the {254 nm|313.18 nm|626 nm} ionization scheme and is further discussed in [26].

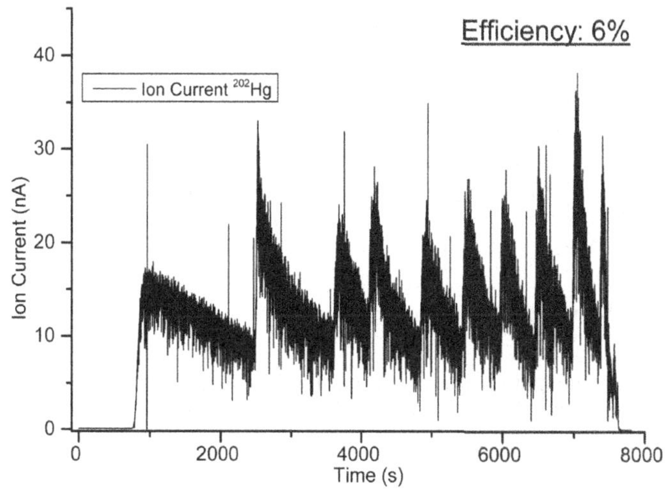

Fig. 7.6 The measured ion current over the period of the efficiency measurement. The laser parameters are presented in Table 7.4. Figure adapted from [26], licensed under CC by 4.0

7.4.5 On-Line Application at ISOLDE

This ionization scheme development enabled the in-source resonance ionization spectroscopy of exotic mercury isotopes. The results of these experiments are presented in Chaps. 8–10 and [31–33].

7.4.6 Conclusions

An optimal RILIS ionization scheme of {254 nm|313.18 nm|532 nm^{NR}} was determined for mercury. An ionization scheme of {254 nm|313.18 nm|795 nm^{RY}}, ionizing via a Rydberg level, corresponded to a factor of four lower ion current at typical laser power levels available at the ISOLDE RILIS for these wavelengths. No autoionizing resonances were observed, including the autoionizing state reported in literature at 88 759.6 cm^{-1}. The efficiency of the new RILIS ionization scheme for mercury of {254 nm|313.18 nm|532 nm^{NR}} was determined to be 6%, a factor of 60 higher than the previously measured efficiency of 0.1% with a scheme of {254 nm|313 nm|626 nm}.

7.5 Radium

The work presented in this section was published in Spectrochimica Acta Part B: Atomic Spectroscopy, as part of a paper titled "*Radium ionization scheme development: The first observed autoionizing states and optical pumping effects in the hot cavity environment*" [34]. The paper is licensed under CC by 4.0.

7.5.1 Motivation

RILIS ionization of radium was requested for an experiment [35, 36] on the CRIS beam line at ISOLDE [37].

7.5.2 Existing Spectroscopic Information

The application of a RILIS for the ionization of radium was reported at TRIUMF, applying a {483 nm|818 nm|532 nm^{NR}} ionization scheme [38]. A factor of three enhancement over the surface ionized background was observed with the pre-irradiated target heated to ≈1000 °C and the tantalum hot cavity with a rhenium foil at the tip heated with a current typically required to heat it to ≈2000 °C. Based on the spectroscopic information listed in [39], suitable first (714 nm) and second (784 nm) step transitions were identified. These had the potential to result in a more convenient ionization scheme to that applied at TRIUMF, avoiding the need to produce 483 nm light which requires a UV pumped dye laser or the operation of a Ti:Sa laser beyond the typical operating range at the ISOLDE RILIS.

7.5.3 Laser Spectroscopy

Since no stable isotope of radium exists, the radium atoms used for this spectroscopy were produced on-line by impinging 1.4 GeV protons on a to UC_x target as described in Sect. 5.1. The target and tantalum hot cavity were maintained at 1950 °C and 2060 °C respectively throughout these measurements. The first and second step transitions were produced using Ti:Sa lasers. A dye laser operated with ethanol solutions of either Rhodamine 6G, Fluorescein 27 or DCM dyes was scanned to search for autoionizing resonances: {714 nm|784 nm|dye scan}.

The scans of the investigated resonant transitions are presented with a schematic of the schemes in Fig. 7.7, with the spectroscopic information presented in Table 7.5. Initially it was believed that three autoionizing states had been discovered (the first to be reported), with ionization schemes of {714 nm|784 nm|558 nm, 581 nm,

Table 7.5 The resonant transitions investigated during the radium ionization scheme development originating from the $7s^2\ ^1S_0$ ground state (*the 615 nm transition is understood to originate from the 7s7p $^3P^o_2$ level). Spectroscopic information on known lines is taken from [39]. Air wavelengths of the new resonances were calculated using the equation of [21]. The uncertainties are dominated by the widths of the autoionizing resonances. Data reproduced from [34]

Transition (cm^{-1})	Upper state config.	Air wavelength (nm)
0 – 13 999.357	7s7p, $^3P^o_1$	714.12167 (5)
13 999.357 – 26 754.01	7s8s, $^3S_1$	783.812 (2)
26 754.01 – 44 686 (4)	new, J = 0, 1, 2	557.51 (13)
26 754.01 – 43 951 (6)	new, J = 0, 1, 2	581.34 (21)
16 688.54* – 32 941.06	$7p^2$, $^3P_2$	615.119 (5)

615 nm}. After unexpected results when the {714 nm|784 nm|615 nm} ionization scheme was tested at the CRIS beam line (using a charge exchange cell to neutralize the ions before re-ionization) and an additional review of the literature, it was determined that the 615 nm line was a known transition. Additionally, it was determined that the 7s7p $^3P^o_2$ state from which the 615 nm transition originates could only be populated in HC-RILIS operation via a decay from the 7s8s $^3S_1$ level, suggesting optical pumping. Therefore the ionization scheme was in fact {714 nm|784 nm|993 nm^{decay}|615 nm|615 nm, 714 nm, 784 nm}.

7.5.4 Timing Investigation

In order to demonstrate optical pumping within the hot cavity environment, the 615 nm transition was delayed with respect to the 714 and 784 nm transitions while monitoring the ion current on a Faraday cup inserted after the dipole mass separator. The results are presented in Fig. 7.8, together with the results from delaying the 558 nm transition in the same manner for comparison.

The insensitivity of the timing synchronization of the 615 nm transition with the first and second steps, can be understood as resulting from the lack of an allowed transition to the atomic ground state from the 7s7p $^3P^o_2$ state. Surprisingly, the {714 nm|784 nm|993 nm^{decay}|615 nm|615 nm, 714 nm, 784 nm} ionization scheme offered a similar factor of three enhancement as the ionization scheme employing the better of the two autoionizing states ({714 nm|784 nm|558 nm^{AI}}).

7.5.5 Saturation Measurement

Saturation measurements of the 714, 784, 558 and 615 nm transitions were undertaken following the methodology presented in Sect. 7.1.3 to further clarify the results

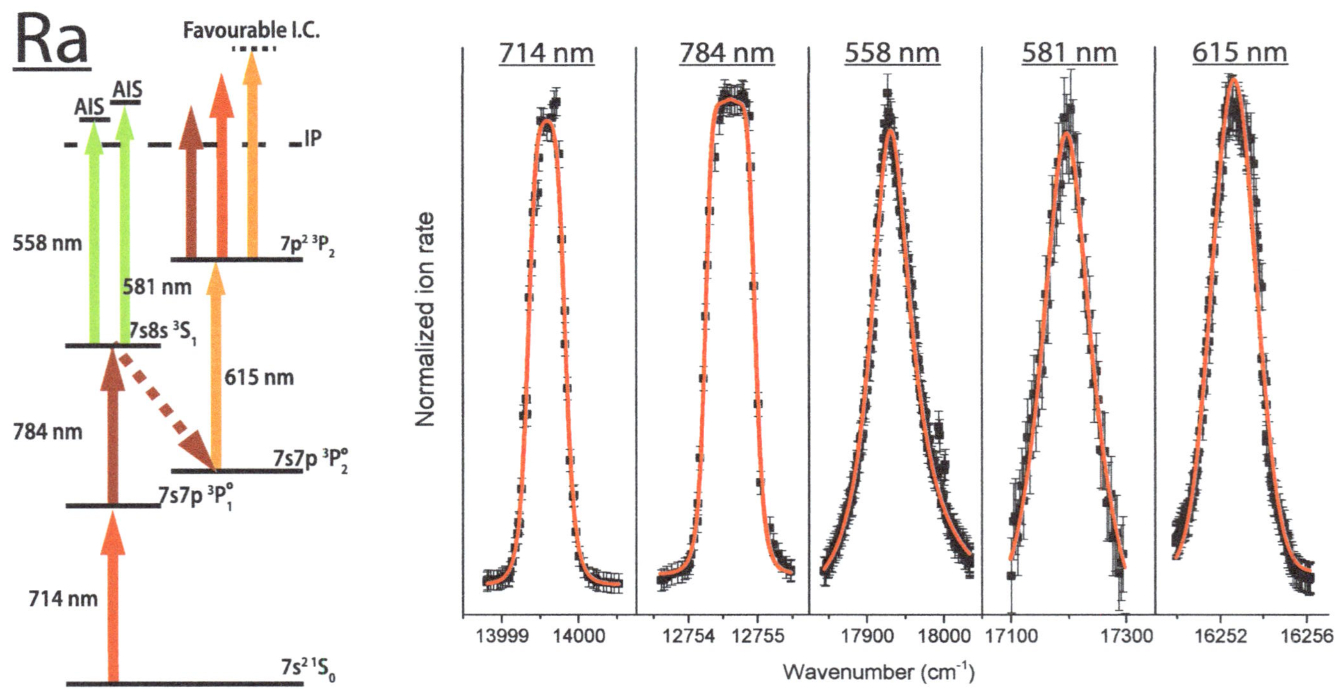

Fig. 7.7 The resonant transitions investigated during the radium ionization scheme development and scans of the resonances. The error bars are estimated based on the precision of the Faraday cup and the method of data acquisition. Figure adapted from [34], licensed under CC by 4.0

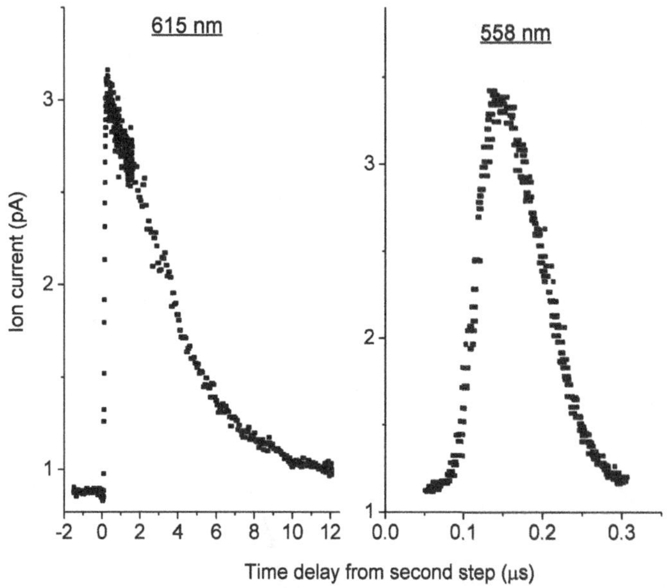

Fig. 7.8 The change in ion current with a variation in the synchronization of the second and third step transitions. An off-set of ≈100 ns with respect to the timing of the 714 and 784 nm transitions is the result of the laser pulse timing being measured at different locations. Figure adapted from [34], licensed under CC by 4.0

and assess the ionization scheme's applicability for on-line operation. The results are presented in Fig. 7.9.

The 714, 784 and 615 nm transitions were found to be deeply saturated. Therefore the {714 nm|784 nm|615 nm} ionization scheme was identified as optimal for on-line RILIS operation. The saturation and ionization efficiency of the 615 nm transition is difficult to understand unless it is capable of reaching a favourable part of the ionization continuum from the $7p^2\ ^3P_2$ excited state. The comparative insensitivity of the 615 nm transition, to delaying it with respect to the first two excitation steps, indicates this transition is also the dominant ionizing step in the {714 nm|784 nm|993 nm^{decay}|615 nm|615 nm, 714 nm, 784 nm} ionization scheme. Thus the ionization scheme can be considered to be {714 nm|784 nm|993 nm^{decay}| 615 nm|615 nm}.

7.5.6 On-Line Application

The {714 nm|784 nm|993 nm^{decay}|615 nm|615 nm} ionization scheme was successfully applied on two occasions to enhance the radium yields for experiments on the CRIS experimental beamline [36].

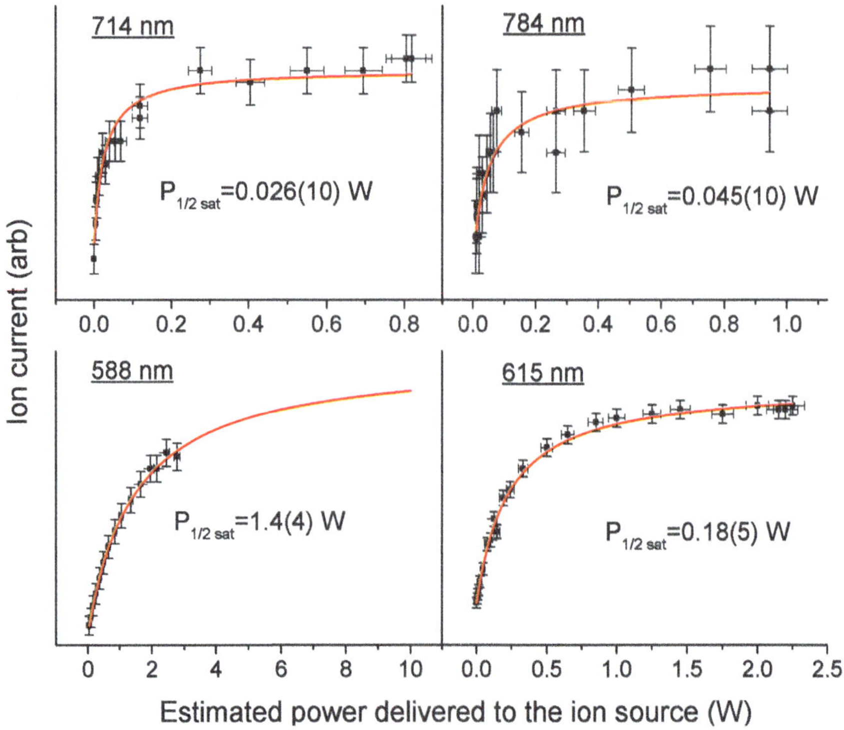

Fig. 7.9 Saturation measurements plotting the laser power delivered to the hot cavity ion source against ion current measured on a Faraday cup for four of the investigated transitions. The uncertainty on individual measurement points is related to the precision of the measurement devices and the method of recording. An additional uncertainty is included for the $P_{1/2sat}$ values, considering a transmission efficiency of the laser light to the ion source between 40–60%. Figure adapted from [34], licensed under CC by 4.0

7.5.7 Conclusion

Two new RILIS ionization schemes for radium have been developed ({714 nm |784 nm|993 nm^{decay}|615 nm|615 nm}, {714 nm|784 nm|588 nm^{AI}}), capable of enhancing the radium yield by a factor of three when applied in a tantalum hot cavity operated at 2060 °C. All three of the transitions were determined to be saturated. During the course of the scheme development, autoionizing states of radium were identified for the first time at 43 951 (6) and 44 686 (4) cm^{-1}. Additionally, optical pumping effects were observed within the hot cavity environment.

7.6 Tellurium

The work presented in this section was published in two papers. The first in Nuclear Instruments and Methods in Physics Research Section A: Accelerators, Spectrometers, Detectors and Associated Equipment, in a paper titled "*Laser resonance ionization scheme development for tellurium and germanium at the dual Ti:Sa-Dye ISOLDE RILIS*" [23]. The second in Hyperfine Interactions, in a paper titled "*RILIS-ionized mercury and tellurium beams at ISOLDE CERN*" [26]. The papers are licensed under CC by 4.0.

7.6.1 Motivation

In addition to ion beam production, the RILIS can also serve as a tool for signal identification. In the majority of cases, blocking or detuning one of the resonant transitions should only influence the element of interest in the ion beam delivered to an experiment. RILIS ionized tellurium is required to enable signal identification for an approved Coulomb excitation experiment investigating ^{116}Te and ^{118}Te [40]. At ISOLDE, tellurium ion beams have thus far been produced using FEBIAD type ion sources, which do not offer the element selective capabilities of a RILIS [41].

7.6.2 Existing Spectroscopic Information

Laser resonance ionization of tellurium was applied previously at ISOLDE in an experiment on the COMPLIS beam line where a two-resonance, three-step ionization scheme of {214 nm|591 nm|1064 nm^{NR}} was applied [42]. The laser atom interaction region had significant differences to the HC-RILIS environment, thus there were no measured efficiencies applicable to this work. A rich continuum of autoionizing states of tellurium was identified previously using flash pyrolysis and other non-laser light sources [43–45]. Additionally, autoionizing resonances have also been investigated at Oak Ridge National Laboratory (private communication Y. Liu).

7.6.3 Laser Spectroscopy

Based on information from the NIST database [3] and Eqs. (7.1)–(7.3), 95% of the atomic population of tellurium lies in the atomic ground state at 2300 K. Thus only schemes involving transitions from the atomic ground state were considered. According to the NIST database [3], the first-step transition from the atomic ground state within the RILIS wavelength range and with the largest photo absorption cross-

Table 7.6 The resonant transitions tested during the tellurium scheme development. Spectroscopic information for known transitions is taken from [3]. *Air wavelengths calculated using the equation of [21]. †Ion current measured with ~24 W of 532 nm light delivered to the ion source for the third, non-resonant, ionization step

Transition (cm^{-1})	Upper state config.	Air wavelength (nm)	Relative ion current
0 – 46 652.738	$5p^3(^4S°)6s\ ^3S_1^°$	214.281	–
46 652.738 – 63 669.944	$5p^3(^4S°)7p\ ^5P_1$	587.478*	0.35†
46 652.738 – 63 921.485	$5p^3(^4S°)7p\ ^3P_2$	578.922	1†
46 652.738 – 63 982.463	$5p^3(^4S°)7p\ ^3P_0$	576.883*	0.24†
46 652.738 – 64 088.997	$5p^3(^4S°)7p\ ^3P_1$	573.360	0.74†
64 088.997 – 75 181.41 (5)	New	901.270*	2.5

section is the 214 nm ($5p^4\ ^3P_2 \longrightarrow 5p^3(^4S°)6s\ ^3S_1^°$) transition. The four second-step transitions corresponding to the highest observed ion signals during the scan, were tested as part of three-step, two-resonance ionization schemes, using 532 nm photons from the Blaze laser for non-resonant ionization. The relative ion currents recorded for each scheme are compared in Table 7.6. A search for autoionizing resonances was conducted by replacing the 532 nm third-step laser with a tunable Ti:Sa laser. A new autoionizing resonance at 901.270 nm was identified as part of a {214 nm|573 nm|901 nm^{AI}} scheme. The spectroscopic investigation is further discussed in [23].

Comparing the measured ion currents, the autoionizing resonance was a factor of 2.5 more efficient than non-resonant ionization with an estimated 24 W delivered to the ion source. The optimal ionization scheme and scans of the 573 nm second-step transition and the autoionizing resonance are presented in Fig. 7.10.

Due to the available time for the ionization scheme development, the saturation of the transitions was verified using a ND (neutral density)=0.6 filter, which had the effect of reducing the transmitted light by a factor of ~4. A negligible effect on the ion current was observed when the filter was inserted, therefore all of the transitions were understood to be saturated. The estimated laser powers delivered to the ion source without a filter inserted is listed in Table 7.7.

7.6.4 *Efficiency*

The efficiency of the {214 nm|573 nm|901 nm^{AI}} ionization scheme was measured following the methodology presented in Sect. 7.1.3. The measurement was made by

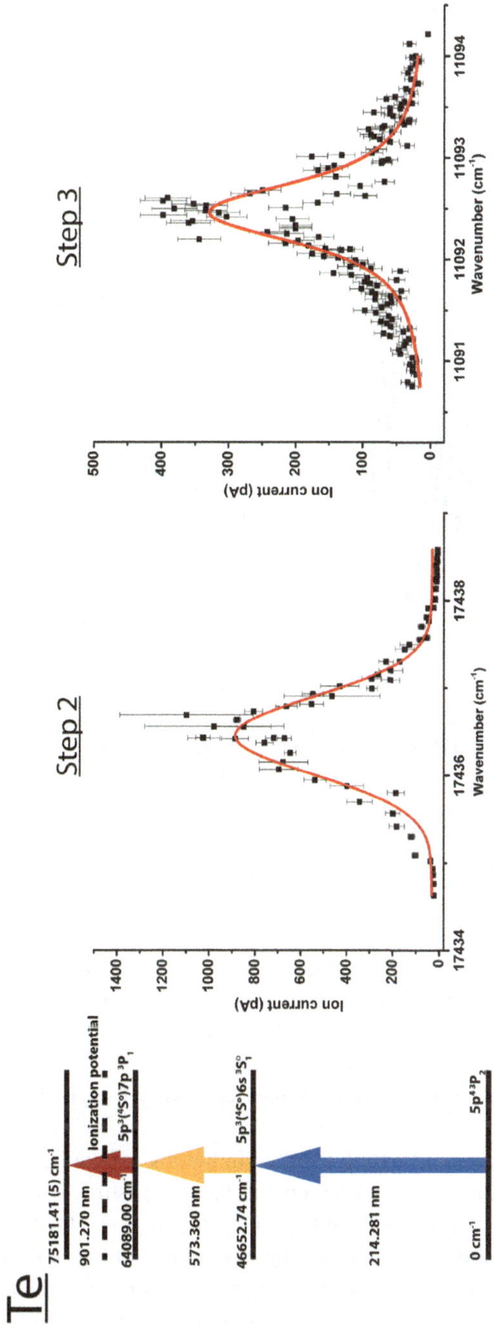

Fig. 7.10 The optimal tellurium ionization scheme and laser scans of the second-step and the third-step autoionizing resonance. Figure reproduced from [23], licensed under CC by 4.0

Table 7.7 The resonant transitions for the ionization scheme originating from the $5p^4\ ^3P_2$ atomic ground state applied during the tellurium ionization scheme efficiency measurement. The given laser power is an estimation of the power delivered to the ion source. Known spectroscopic information taken from [23]

Transition (cm^{-1})	Air wavelength (nm)	Laser power (mW)
0 – 46 652.738	214.281	13
46 652.738 – 64 088.997	573.360	4 000
64 088.997 – 75 181.41 (20)	901.270 (17)	1 300

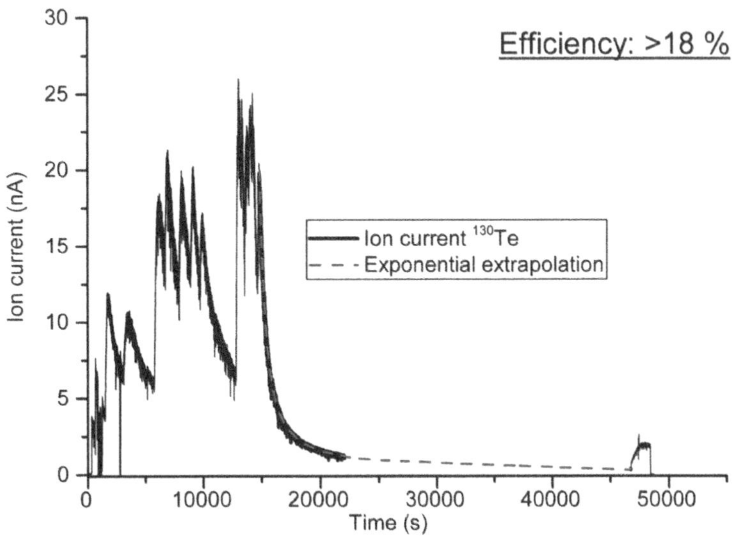

Fig. 7.11 The measured ion current over the period of the efficiency measurement, together with a schematic of the ionization scheme that was applied. The laser parameters are presented in Table 7.7. Figure adapted from [26], licensed under CC by 4.0

monitoring the ion current of ^{130}Te, the isotope of tellurium with the highest natural abundance (34.08 (62)% [46]). The laser parameters during the tellurium efficiency measurements are presented in Table 7.7.

A calibrated sample of 1 000 nAh of tellurium (≈340 nAh of ^{130}Te) was evaporated, with a Faraday cup measurement recorded every 1 s. The results are presented in Fig. 7.11.

A data acquisition malfunction resulted in the loss of the Faraday cup measurements towards the end of the measurement. However, sufficient data was recorded to enable a reliable extrapolation of the ion current to cover the missing data points using an exponential decay function. This is depicted in Fig. 7.11 as a dashed (red) line. The sample was not fully evaporated in the time available indicating a significant amount of tellurium remaining in the system, this is further discussed in [26]. An average signal of 4.8 nA was recorded over a period of 12.98 h, which corresponds to an efficiency of 18%. Due to the tellurium remaining in the system, this is considered to be a lower limit for the efficiency of the ionization scheme.

7.6.5 Conclusion

A new three-step, three-resonance RILIS ionization scheme for tellurium was developed applying a final step transition to a newly identified autoionizing state. The efficiency of the ionization scheme was determined to be >18%, with a significant amount of tellurium remaining in the system at the end of the measurement.

References

1. Fedosseev VN, Marsh BA, Fedorov DV, Köster U, Tengborn E (2006) Ionization scheme development at the ISOLDE RILIS. Hyperfine Interact 162(1–4):15–27. https://doi.org/10.1007/s10751-005-9204-2
2. Smith PL, Claas H, Esmond JR, Kurucz RL (2015) Atomic spectral line database. http://www.pmp.uni-hannover.de/cgi-bin/ssi/test/kurucz/sekur.html. Accessed 28 Sept 2015
3. Kramida A, Ralchenko Yu, Reader J, NIST ASD Team (2015) NIST atomic spectra database (ver. 5.2). http://physics.nist.gov/asd. Accessed 22 July 2016
4. Saloman EB (1990) A resonance ionization spectroscopy/resonance ionization mass spectrometry data service. I—Data sheets for As, B, Cd, C, Ge, Au, Fe, Pb, Si and Zn. Spectrochim Acta Part B: At Spectrosc 45(1–2):37–83. https://doi.org/10.1016/0584-8547(90)80080-3
5. Saloman EB (1991) A resonance ionization spectroscopy/resonance ionization mass spectrometry data service. II—Data sheets for Al, Ca, Cs, Cr, Co, Cu, Kr, Mg, Hg and Ni. Spectrochim Acta Part B: At Spectrosc 46(3):319–378. https://doi.org/10.1016/0584-8547(91)80035-2
6. Saloman EB (1992) A resonance ionization spectroscopy/resonance ionization mass spectrometry data service. Spectrochim Acta Part B: At Spectrosc 47(4):517–543. https://doi.org/10.1016/0584-8547(92)80044-H
7. Saloman EB (1993) A resonance ionization spectroscopy/resonance ionization mass spectrometry data service. Spectrochim Acta Part B: At Spectrosc 48(9):1139–1203. https://doi.org/10.1016/0584-8547(93)80104-3
8. Saloman EB (1994) A resonance ionization spectroscopy/resonance ionization mass spectrometry data service. Spectrochim Acta Part B: At Spectrosc 49(3):251–281. https://doi.org/10.1016/0584-8547(94)80023-5
9. Saloman EB (2006) Wavelengths, energy level classifications, and energy levels for the spectrum of neutral mercury. J Phys Chem Ref Data 35(4):1519–1548. https://doi.org/10.1063/1.2204960
10. Saloman EB (2012) Energy levels and observed spectral lines of neutral and singly ionized chromium, Cr I and Cr II. J Phys Chem Ref Data 41(4):043103. https://doi.org/10.1063/1.4754694
11. Fedosseev VN, Yusupov AK (1984) Effect of polarization of laser emission on the multistage photoionization yield of ytterbium isotopes. Opt Spectrosc (English translation of Optika i Spektroskopiya) 57:905–908
12. Rossel RE, Fedosseev VN, Marsh BA, Richter D, Rothe S, Wendt KDA (2013) Data acquisition, remote control and equipment monitoring for ISOLDE RILIS. Nucl Instrum Methods Phys Res Sect B: Beam Interact Mater At 317:557–560. https://doi.org/10.1016/j.nimb.2013.05.048

13. Rossel RE (2015) A distributed monitoring and control system for the laser ion source RILIS at CERN-ISOLDE. PhD thesis, Hochschule RheinMain. https://doi.org/10.17181/CERN.L5N9.2GCS
14. Rothe S (2012) An all-solid state laser system for the laser ion source RILIS and in-source laser spectroscopy of astatine at ISOLDE, CERN. Doctoral thesis, J.G. University of Mainz. https://doi.org/10.17181/CERN.0HNZ.7Z6X
15. Fink DA (2014) Improving the selectivity of the ISOLDE resonance ionization laser ion source and in-source laser spectroscopy of polonium. Doctoral thesis, Heidelberg University. https://doi.org/10.17181/CERN.R8QG.WACQ
16. Liu Y, Baktash C, Beene JR, Havener CC, Krause HF, Schultz DR, Stracener DW, Vane CR, Geppert Ch, Kessler T, Wies K, Wendt K (2011) Time profiles of ions produced in a hot-cavity resonant ionization laser ion source. Nucl Instrum Methods Phys Res Sect B: Beam Interact Mater At 269(23):2771–2780. https://doi.org/10.1016/j.nimb.2011.08.009
17. Day Goodacre T, Chrysalidis K, Fedorov DV, Fedosseev VN, Marsh BA, Molkanov PL, Rossel RE, Rothe S, Seiffert C (2017) The identification of autoionizing states of atomic chromium for the resonance ionization laser ion source of the ISOLDE radioactive ion beam facility. Spectrochim Acta Part B: At Spectrosc 129:58–63. https://doi.org/10.1016/j.sab.2017.01.001, http://linkinghub.elsevier.com/retrieve/pii/S0584854717300095
18. Kreim S, Beck D, Blaum K, Böhm Ch, Borgmann Ch, Breitenfeldt M, Cakirli RB, Herfurth F, Kowalska M, Litvinov Y, Lunney D, Manea V, Naimi S, Neidherr D, Rosenbusch M, Schweikhard L, Stanja J, Stora T, Wienholtz F, Wolf RN, Zuber K (2011) Seeking the purported magic number N = 32 with high-precision mass spectrometry. Technical report, CERN-INTC, Geneva. https://doi.org/10.17181/CERN.P6YL.2WO1
19. Stora T (2013) Recent developments of target and ion sources to produce ISOL beams. Nucl Instrum Methods Phys Res Sect B: Beam Interact Mater At 317:402–410. https://doi.org/10.1016/j.nimb.2013.07.024
20. Kirchner R (1981) Progress in ion source development for on-line separators. Nucl Instrum Methods Phys Res 186(1–2):275–293. https://doi.org/10.1016/0029-554X(81)90916-2
21. Birch KP, Downs MJ (1993) An updated Edlén equation for the refractive index of air. Metrologia 30(3):155–162. https://doi.org/10.1088/0026-1394/30/3/004
22. Mougeot M, Atanasov D, Blaum K, Chrysalidis K, Day Goodacre T, Fedorov D, Fedosseev V, George S, Herfurth F, Holt JD, Lunney D, Manea V, Marsh B, Neidherr D, Rosenbusch M, Rothe S, Schweikhard L, Schwenk A, Seiffert C, Simonis J, Stroberg SR, Welker A, Wienholtz F, Wolf RN, Zuber K (2018) Precision mass measurements of 58-63Cr: nuclear collectivity towards the N=40 island of inversion. Phys Rev Lett 120(23):232501, 2018. https://doi.org/10.1103/PhysRevLett.120.232501
23. Day Goodacre T, Fedorov D, Fedosseev VN, Forster L, Marsh BA, Rossel RE, Rothe S, Veinhard M (2016) Laser resonance ionization scheme development for tellurium and germanium at the dual Ti:Sa–Dye ISOLDE RILIS. Nucl Instrum Methods Phys Res Sect A: Accel Spectrom Detect Assoc Equip 830:510–514. https://doi.org/10.1016/j.nima.2015.10.066
24. Nacher E, Briz JA, Carmona M, Illana A, Jungclaus A, Perea A, Pesudo V, Ribeiro G, Sánchez-del Río J, Sarriguren P, Taprogge J, Tengblad O, Domingo C, Algora A, Agramunt J, Giubrone G, Guadilla V, Montaner A, Orrigo SEA, Rubio B, Taín JL, Valencia E, José J, Parikh A, Fraile LM, Marroquín I, Moreno O, Olaizola B, Paziy V, Udías JM, Vedia V, Borge MJG, Day Goodacre T, Fedosseev V, Marsh B, Rapisarda E, Stora T, Gelletly W, Regan P, Podolyák Z, Rice S, Orlandi R (2013) Beta decay of the N=Z, rp-process waiting points: 64Ge, 68Se and the N=Z+2: 66Ge, 70Se for accurate stellar weak-decay rates. CERN-INTC 11. http://cds.cern.ch/record/1551738/files/INTC-P-374.pdf
25. Liu Y, Baktash C, Beene JR, Bilheux HZ, Havener CC, Krause HF, Schultz DR, Stracener DW, Vane CR, Brück K, Geppert Ch, Kessler T, Wendt K (2006) Laser ion source tests at the HRIBF on stable Sn, Ge and Ni isotopes. Nucl Instrum Methods Phys Res Sect B: Beam Interact Mater At 243(2):442–452. https://doi.org/10.1016/j.nimb.2005.09.014, http://linkinghub.elsevier.com/retrieve/pii/S0168583X05017623

26. Day Goodacre T, Billowes J, Chrysalidis K, Fedorov DV, Fedosseev VN, Marsh BA, Molkanov PL, Rossel RE, Rothe S, Seiffert C, Wendt KDA (2017) RILIS-ionized mercury and tellurium beams at ISOLDE CERN. Hyperfine Interact 238(1):41. https://doi.org/10.1007/s10751-017-1398-6
27. Podshivalov AA, Matveev OI, Smith BW, Winefordner JD (1999) A novel and efficient excitation- and ionization-scheme for laser resonance ionization of mercury. Spectrochim Acta Part B: At Spectrosc 54(13):1793–1799. https://doi.org/10.1016/S0584-8547(99)00112-3
28. Fedosseev VN, Berg L-E, Lebas N, Launila OJ, Lindroos M, Losito R, Marsh BA, Österdahl FK, Pauchard T, Tranströmer G, Vannesjö J (2008) ISOLDE RILIS: new beams, new facilities. Nucl Instrum Methods Phys Res Sect B: Beam Interact Mater At 266(19–20):4378–4382. https://doi.org/10.1016/j.nimb.2008.05.038
29. Sundell S, Ravn H (1992) Ion source with combined cathode and transfer line heating. Nucl Instrum Methods Phys Res Sect B: Beam Interact Mater At 70(1–4):160–164. https://doi.org/10.1016/0168-583X(92)95926-I
30. Marsh B, Day Goodacre T, Fink D, Rothe S, Seliverstov M, Imai N, Sjodin M, Rossel R (2013) Suitability test of a high beam quality Nd:YVO4 industrial laser for the ISOLDE RILIS installation. Technical report, CERN, Geneva. https://doi.org/10.17181/CERN.F65D.P3NR
31. Marsh BA, Day Goodacre T, Sels S, Tsunoda Y, Andel B, Andreyev AN, Althubiti NA, Atanasov D, Barzakh AE, Billowes J, Blaum K, Cocolios TE, Cubiss JG, Dobaczewski J, Farooq-Smith GJ, Fedorov DV, Fedosseev VN, Flanagan KT, Gaffney LP, Ghys L, Huyse M, Kreim S, Lunney D, Lynch KM, Manea V, Martinez Palenzuela Y, Molkanov PL, Otsuka T, Pastore A, Rosenbusch M, Rossel RE, Rothe S, Schweikhard L, Seliverstov MD, Spagnoletti P, Van Beveren C, Van Duppen P, Veinhard M, Verstraelen E, Welker A, Wendt K, Wienholtz F, Wolf RN, Zadvornaya A, Zuber K (2018) Characterization of the shape-staggering effect in mercury nuclei. Nat Phys. https://doi.org/10.1038/s41567-018-0292-8
32. Sels S, Day Goodacre T, Marsh BA, Pastore A, Ryssens W, Tsunoda Y, Althubiti N, Andel B, Andreyev AN, Atanasov D, Barzakh AE, Bender M, Billowes J, Blaum K, Cocolios TE, Cubiss JG, Dobaczewski J, Farooq-Smith GJ, Fedorov DV, Fedosseev VN, Flanagan KT, Gaffney LP, Ghys L, Heenen P-H, Huyse M, Kreim S, Lunney D, Lynch KM, Manea V, Martinez Palenzuela Y, Medonca TM, Molkanov PL, Otsuka T, Ramos JP, Rossel RE, Rothe S, Schweikhard L, Seliverstov MD, Spagnoletti P, Van Beveren C, Van Duppen P, Veinhard M, Verstraelen E, Welker A, Wendt K, Wienholtz F, Wolf RN, Zadvornaya A (2019) Shape staggering of midshell mercury isotopes from in-source laser spectroscopy compared with density-functional-theory and Monte Carlo shell-model calculations. Phys Rev C 99(4):044306. https://doi.org/10.1103/PhysRevC.99.044306
33. Day Goodacre T, Afanasjev AV, Barzakh AE, Marsh BA, Sels S, Ring P, Nakada H, Andreyev AN, Van Duppen P, Althubiti NA, Andel B, Atanasov D, Billowes J, Blaum K, Cocolios TE, Cubiss JG, Farooq-Smith GJ, Fedorov DV, Fedosseev VN, Flanagan KT, Gaffney LP, Ghys L, Huyse M, Kreim S, Lunney D, Lynch KM, Manea V, Martinez Palenzuela Y, Molkanov PL, Rosenbusch M, Rossel RE, Rothe S, Schweikhard L, Seliverstov MD, Spagnoletti P, Van Beveren C, Veinhard M, Verstraelen E, Welker A, Wendt K, Wienholtz F, Wolf RN, Zadvornaya A, Zuber K (2021) Laser spectroscopy of neutron-rich 207,208Hg isotopes: illuminating the kink and odd-even staggering in charge radii across the N=126 shell closure. Phys Rev Lett 126(3):032502. https://doi.org/10.1103/PhysRevLett.126.032502
34. Day Goodacre T, Billowes J, Binnersley CL, Bissell ML, Chrysalidis K, Cocolios TE, de Groote RP, Farooq-Smith GJ, Fedorov DV, Fedosseev VN, Flanagan KT, Franchoo S, Garcia Ruiz RF, Gins W, Heinke R, Koszorús Á, Lynch KM, Marsh BA, Molkanov PL, Naubereit P, Neyens G, Ricketts CM, Rothe S, Seiffert C, Seliverstov MD, Stroke HH, Studer D, Vernon AR, Wilkins SG, Wendt KDA, Yang XF (2018) Radium ionization scheme development: the first observed autoionizing states and optical pumping effects in the hot cavity environment. Spectrochim Acta Part B: At Spectrosc 150:99–104. https://doi.org/10.1016/j.sab.2018.10.002
35. Lynch KM, Billowes J, Bissell ML, Budinčević I, Cocolios TE, Day Goodacre T, de Groote RP, Fedosseev VN, Flanagan KT, Franchoo S, Garcia Ruiz RF, Heylen H, Kron T, Marsh BA, Neyens G, Rossel RE, Rothe S, Trashnov I, Stroke HH, Wendt KDA (2014) Collinear resonance

ionization spectroscopy of radium ions. Technical report, CERN-INTC, Geneva. https://cds.cern.ch/record/1704513
36. Lynch KM, Wilkins SG, Billowes J, Binnersley CL, Bissell ML, Chrysalidis K, Cocolios TE, Day Goodacre T, de Groote RP, Farooq-Smith GJ, Fedorov DV, Fedosseev VN, Flanagan KT, Franchoo S, Garcia Ruiz RF, Gins W, Heinke R, Koszorús Á, Marsh BA, Molkanov PL, Naubereit P, Neyens G, Ricketts CM, Rothe S, Seiffert C, Seliverstov MD, Stroke HH, Studer D, Vernon AR, Wendt KDA, Yang XF (2018) Laser-spectroscopy studies of the nuclear structure of neutron-rich radium. Phys Rev C 97(2):024309. https://doi.org/10.1103/PhysRevC.97.024309
37. Cocolios TE, Al Suradi HH, Billowes J, Budinčević I, de Groote RP, De Schepper S, Fedosseev VN, Flanagan KT, Franchoo S, Garcia Ruiz RF, Heylen H, Le Blanc F, Lynch KM, Marsh BA, Mason PJR, Neyens G, Papuga J, Procter TJ, Rajabali MM, Rossel RE, Rothe S, Simpson GS, Smith AJ, Strashnov I, Stroke HH, Verney D, Walker PM, Wendt KDA, Wood RT (2013) The collinear resonance ionization spectroscopy (CRIS) experimental setup at CERN-ISOLDE. Nucl Instrum Methods Phys Res Sect B: Beam Interact Mater At 317:565–569. https://doi.org/10.1016/j.nimb.2013.05.088
38. Raeder S, Lassen J, Heggen H, Teigelhöfer A (2014) In-source spectroscopy on astatine and radium for resonant laser ionization. Hyperfine Interact 227(1–3):77–83. https://doi.org/10.1007/s10751-014-1040-9
39. Dammalapati U, Jungmann K, Willmann L (2016) Compilation of spectroscopic data of radium (Ra I and Ra II). J Phys Chem Ref Data 45(1):013101. https://doi.org/10.1063/1.4940416
40. Ahn T, Al-Azri H, Bloch T, Butler PA, Bree N, Bäck T, Bönig S, Cederkäll J, Cederwall B, Darby IG, Diriken J, O'Donnell D, Fahlander C, Gaffney LP, Grahn T, Hadinia B, Huyse M, Jenkins DG, Johnson A, Joshi P, Joss DT, Julin R, Kröll T, Leske J, Nara Singh BS, Nicholls A, Page RD, Pakarinen J, Paul ES, Pietralla N, Rahkila P, Rapisarda E, Sandzelius M, Scheck M, Simpson J, Smith JF, Wadsworth R, Van Duppen P, Voulot D, Wenander F, Werner V (2011) Coulomb excitation of 116Te and 118Te: a study of collectivity above the Z=50 shell gap. CERN-INTC 11. https://doi.org/10.17181/CERN.LT8P.T00C
41. Köster U, Bergmann UC, Carminati D, Catherall R, Cederkäll J, Correia JG, Crepieux B, Dietrich M, Elder K, Fedoseyev VN, Fraile L, Franchoo S, Fynbo H, Georg U, Giles T, Joinet A, Jonsson OC, Kirchner R, Lau Ch, Lettry J, Maier HJ, Mishin VI, Oinonen M, Peräjärvi K, Ravn HL, Rinaldi T, Santana-Leitner M, Wahl U, Weissman L (2003) Oxide fiber targets at ISOLDE. Nucl Instrum Methods Phys Res Sect B: Beam Interact Mater At 204:303–313. https://doi.org/10.1016/S0168-583X(03)00505-6
42. Sifi R, Le Blanc F, Barré N, Cabaret L, Crawford J, Ducourtieux M, Essabaa S, Genevey J, Huber G, Kowalska M, Lau C, Lee JKP, Le Scornet G, Oms J, Pinard J, Roussière B, Sauvage J, Seliverstov M, Stroke H (2006) Laser spectroscopy measurements of neutron-rich tellurium isotopes by COMPLIS. Hyperfine Interact 171(1–3):173–179. https://doi.org/10.1007/s10751-006-9505-0
43. Berkowitz J, Batson CH, Goodman GL (1981) Photoionization of atomic iodine and atomic tellurium. Phys Rev A 24(1):149–160. https://doi.org/10.1103/PhysRevA.24.149
44. Cantú AM, Mazzoni M, Joshi YN (1983) Rydberg series in the absorption spectrum of Te I limiting on 5s2 5p3 4S3/2 0 ionization limit. Phys Scr 27(4):261–266. https://doi.org/10.1088/0031-8949/27/4/007, http://stacks.iop.org/1402-4896/27/i=4/a=007?key=crossref.a464f255f9aeef8536b2d1ebd00c781d
45. Mazzoni M, Cantu AM, Joshi YN (1983) Photoionisation spectrum of Te I: autoionising series and relative cross section. J Phys B: At Mol Phys 16(17):3183–3190. https://doi.org/10.1088/0022-3700/16/17/011, http://stacks.iop.org/0022-3700/16/i=17/a=011?key=crossref.3cc91adb3ba49ac72b39dcab45722ee4
46. de Laeter JR, Böhlke JK, De Bièvre P, Hidaka H, Peiser HS, Rosman KJR, Taylor PDP (2003) Atomic weights of the elements. Review 2000 (IUPAC technical report). Pure Appl Chem 75(6). https://doi.org/10.1351/pac200375060683

Chapter 8
In-Source Resonance Ionization Spectroscopy of Mercury

Abstract This chapter presents the results of the in-source spectroscopy of exotic mercury isotopes. The remeasured isotopes are compared with literature results and the possible sources of experimental uncertainty are explored.

8.1 Introduction

The in-source RIS experiment [1] discussed in the following chapters took place at the ISOLDE facility described in Sect. 5.1. Some of the results have already been published in combination with theoretical calculations [2–4]. Mercury isotopes were produced by impinging a pulsed 1.4 GeV proton "driver beam" on to a molten lead target. A staggered proton pulse substructure was used to maximize the release of short-lived isotopes from the molten target [5, 6]. For molten targets, the limit on the number of protons per pulse is reduced from 3.5×10^{13} to 1×10^{13} in order to preserve the target containers [7]. Therefore, the achievable proton current is limited by the number of proton pulses available from the CERN proton sharing supercycle (typically 35–40% of the CERN total), rather than by limits set by radiation protection requirements. In order to maximize yields for this experiment, agreements were made with other CERN facilities that enabled up to 68% of the CERN proton pulses being made available to ISOLDE.

This experiment was the first on-line application of the VADLIS ion source [8]: coupling the ISOLDE RILIS [9] with the VADIS [10], ISOLDE's variant of the FEBIAD type arc discharge ion source [11–13]. The developments leading to this capability are presented in Chap. 6. The VADLIS developments enabled the RILIS to be coupled to a molten metal target for the first time. The specific benefits for this experiment are discussed in Sect. 8.2.

The first step ($5d^{10}6s^{2\,1}S_0 \rightarrow 5d^{10}6s6p\,^3P_1^\circ$) transition of the three step ionization scheme, described in Sect. 7.4, was chosen as the spectroscopic transition due to its known sensitivity to variations in nuclear structure [14]. The 254 nm radiation required for this transition was generated by 3ω frequency conversion of the output of

T. Day Goodacre, *Applied Laser Spectroscopy for Nuclear Physics*, Springer Theses,
https://doi.org/10.1007/978-3-030-73889-1_8

a Ti:Sa laser, operating in the dual etalon scanning mode described in Sect. 5.3.1. The laser was scanned in a stepwise manner and at each frequency step the ion rate was recorded either via direct ion measurement or α-decay spectroscopy. The technique is summarized in Sects. 4.2.2 and 5.3.

The isotope ^{198}Hg was chosen as the reference isotope and all of the isotope shifts were determined relative to it. Consequently, measurements of ^{198}Hg were repeated periodically during the experiment. The ion rate was determined with an ISOLDE Faraday cup for the first nine of the 17 reference measurements of ^{198}Hg. The ISOLTRAP MR-ToF MS [15] (Sect. 5.3.2) was used to determine the mercury ion rate for the remaining eight scans of ^{198}Hg. The scope of the experiment, including the number of successful scans of each isotope and the detection method employed is presented in Table 8.1.

The Windmill detector [17] (Sect. 5.3.3) was used to determine the ion rate of isotopes A $\leq$ 182. Gating on the α-decay energy enabled isotope specific detection. The MR-ToF MS was used to determine the ion rates of isotopes with A $\geq$ 183, applying timing gates in order to separate mercury specific ion signals from isobaric contamination.

Table 8.1 The number of successful laser scans and the method of determining the laser frequency dependent ion rate for the measured isotopes. The Windmill detector determined the ion rate via the detection of α-decays, while the detector downstream of the MR-ToF MS and the Faraday cup determined it by detecting the charge of the incident ions. Half-lives taken from [16]

Isotope (A)	Half-life (time)	Number of successful scans	Method of ion rate determination
177	127.3 (18) ms	2	Windmill
178	266.5 (24) ms	3	Windmill
179	1.05 (3) s	4	Windmill
180	2.59 (1) s	4	Windmill
181	3.6 (1) s	3	Windmill
182	10.83 (6) s	2	Windmill
183	9.4 (7) s	2	MR-ToF MS
184	30.87 (26) s	3	MR-ToF MS
185g	49.1 (10) s	3	MR-ToF MS
185m	21.6 (15) s	3	MR-ToF MS
198	Stable	9	Faraday cup
198	Stable	8	MR-ToF MS
202	Stable	2	MR-ToF MS
203	46.613 (18) days	2	MR-ToF MS
206	8.32 (7) min	3	MR-ToF MS
207	2.9 (2) min	3	MR-ToF MS
208	42 (5) min	5	MR-ToF MS

8.2 On-Line Application of the VADLIS Ion Source

Measurement of the neutron rich mercury isotopes, beyond the N=126 shell closure, would not have been possible with a UC_x target due to the francium (Z=87) contamination on masses A=206–208. The ISOLTRAP MR-ToF MS is typically capable of separating the isotopes of interest from isobaric contamination, however, there is a limit to the maximum number of ions that can be injected into the MR-ToF MS. Consequently, there is a limit on the ratio of the isotope of interest to the isobaric contamination. Radioactive ion beams of mercury produced in a molten lead (Z=82) target do not suffer from francium contamination. However, molten metal targets are currently incompatible with the surface ion sources typically used as the laser-atom interaction region at ISOLDE. This problem was overcome by the coupling of the ISOLDE RILIS with the VADIS ion source, described in Chap. 5, that is typically used with molten metal targets at ISOLDE.

The VADIS ion source was operated in the new RILIS mode of operation, presented in Chap. 6, to prevent arc discharge ionized stable lead from overwhelming the MR-ToF MS detection system. Operating in RILIS mode reduced the stable lead background by seven orders of magnitude, this can be observed in Fig. 8.1.

Figure 8.1 also demonstrates the necessity of the MR-ToF MS to determine the ion rate. Without the isobaric separation of surface ionized contaminants still present

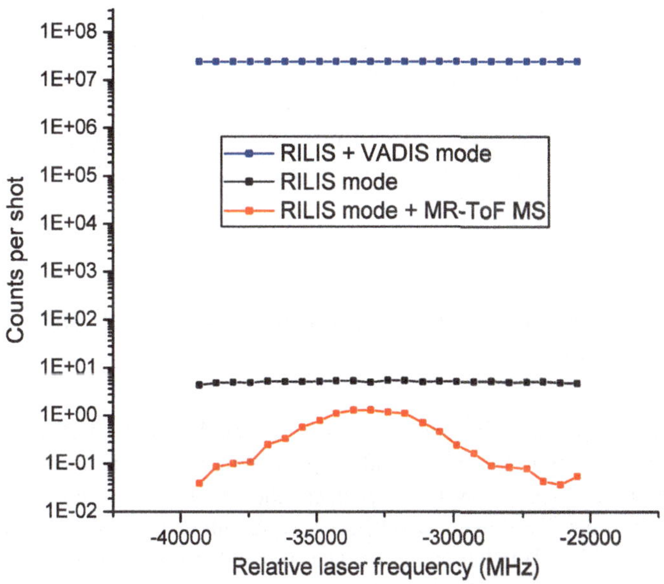

Fig. 8.1 Ion rates on mass A = 208 at the detector down stream of the MR-ToF MS, both directly measured and estimated (the rate for RILIS + VADIS mode was calculated based on Faraday cup measurements)

in the ion beam, determination of the ^{208}Hg related signal would have been close to impossible.

8.3 Fitting Procedure and $\delta\langle r^2\rangle^{198,A}$ Extraction

The laser scans were fitted to extract the centre of gravity (CG) of the HFS, the hyperfine A factor and the hyperfine B factor using a program written in C by A. Barzakh for the Origin 2016 graphing and analysis software. The program considers the resonances to have Voigt profiles and includes 11 parameters to account for the nuclear spin, CG, hyperfine A factor, hyperfine B factor, Lorentzian contribution, Gaussian contribution, amplitude, peak amplitude ratios, a constant and a linearly varying background. The program was applied within the Origin 2016 software environment, with a chi-squared minimization [18] performed by a Levenberg–Marquardt algorithm [19, 20].

The mean square charge radii $\delta\langle r^2\rangle^{198,A}$ was determined using the F factor $F_{254} = -55.36$ GHz/fm^2 obtained by ab initio MCDF calculations by Torbohm et al. [21]. The constant k from the Seltzer moment (Eq. (3.6)) was also taken from [21] as $k = 0.938$. This is in close agreement with the determination of k from the coefficients of [22] ($k = 0.935$). The applied values are the same as those used by Ulm et al. [23]. However, no attempt was made to scale F_{254} according to the mass number as was done in [23]. The specific mass shift coefficient S from Eq. (3.3) was considered to be $S = 0$ [24], the normal mass shift contribution was calculated for each isotope.

In order to enable a comparison of the $\delta\nu^{198,A}$ for ^{182}Hg between this work and literature values, the isotope shift and uncertainty for the literature value of ^{182}Hg was determined from the reported $\delta\nu^{A,A'}$ in the 546 nm ($5d^{10}6s6p\ ^3P_2^\circ \rightarrow 5d^{10}6s7s\ ^3S_1$) line using the F factor conversion $\kappa = F_{546}/F_{254} = -0.1707(12)$ determined in [23].

8.4 Data Analysis and the Determination of Experimental Uncertainties

The ion rate at each laser frequency step was extracted from the MR-ToF MS mass spectrometry data by F. Wienholtz and from the Windmill decay spectroscopy data by S. Sels and E. Verstraelen. Following this, the count-rates were associated with the frequencies measured using two RILIS wavelength meters (WM): a Highfinesse WS7 and a LM-007 ATOS lambdameter. The laser and proton driver beam parameters were monitored continuously throughout the experiment and measurements were associated with each laser frequency step. These data were used to determine the stability of the WMs, the laser power, the laser pulse timing synchronization and the proton driver beam during individual laser scans and over the period of the experiment. The corrected and uncorrected count rates were plotted against the laser frequencies and the plots were fitted using the program described in Sect. 8.3, the results were compared and the optimal choice and application of the CG of the ^{198}Hg reference value was investigated. A summary of the investigations is given below, the full analysis is detailed in the original version of this thesis published on the CERN document server (CDS) [25].

- WM comparison. It was determined that there was a drift in the ATOS WM during the period of the experiment. Additionally there appears to have been an inconsistent illumination of the ATOS interferometers during individual laser scans. Consequently, all of the analysis proceeded using the frequencies measured with the WS7 WM.
- Three options for the application of the ^{198}Hg reference CG were investigated:
  - A global average of the CGs of all of the laser scans of ^{198}Hg.
  - An average of the CGs of the ^{198}Hg laser scans where the MR-ToF MS was used to determine the laser frequency dependent ion rate. This option was considered due to concerns about the effect on the CG of the multiple mercury isotopes present in the ion beam measured with the Faraday cup. This can be observed in the ^{198}Hg (198 FC) scan presented in Fig. 8.2.
  - Interpolating between reference measurements to determine a ^{198}Hg reference value specific to the time of each laser scan.

  It was determined that taking a global average resulted in the closest agreement with literature and the smallest standard deviation between scans of the same element. This value was therefore used for the data analysis.
- An assessment of the saturation of the spectroscopic transition and the effect of correcting the ion rates for variations in the laser power was undertaken. The optimal results were obtained without power correction, indicating the saturation of the spectroscopic transition. This mistake was a consequence of the saturation measurement taking place before the thick etalon was installed to reduce the linewidth of the Ti:Sa laser. A factor of $\approx$6 narrower laser linewidth following the installation of the thick etalon led to a different saturation threshold. This can be understood from Eq. (4.1). Nevertheless, all of the HFS components are well resolved other than for ^{185g,m}Hg. Therefore the saturation does not appear to have been significantly detrimental to the experiment.
- The laser pulse timing remained synchronized during all of the laser scans other than the first scan of ^{179}Hg. The values extracted from this scan were omitted from the final results.
- Large fluctuations (>50%) in the current of the proton driver beam were identified during the laser scans of ^{184}Hg. Surprisingly, this did not appear to affect the measurements of this isotope. This indicates the lead target was in an $\approx$solid state during this period of the experiment, thus significantly increasing the release time of isotopes created by the impinging protons.

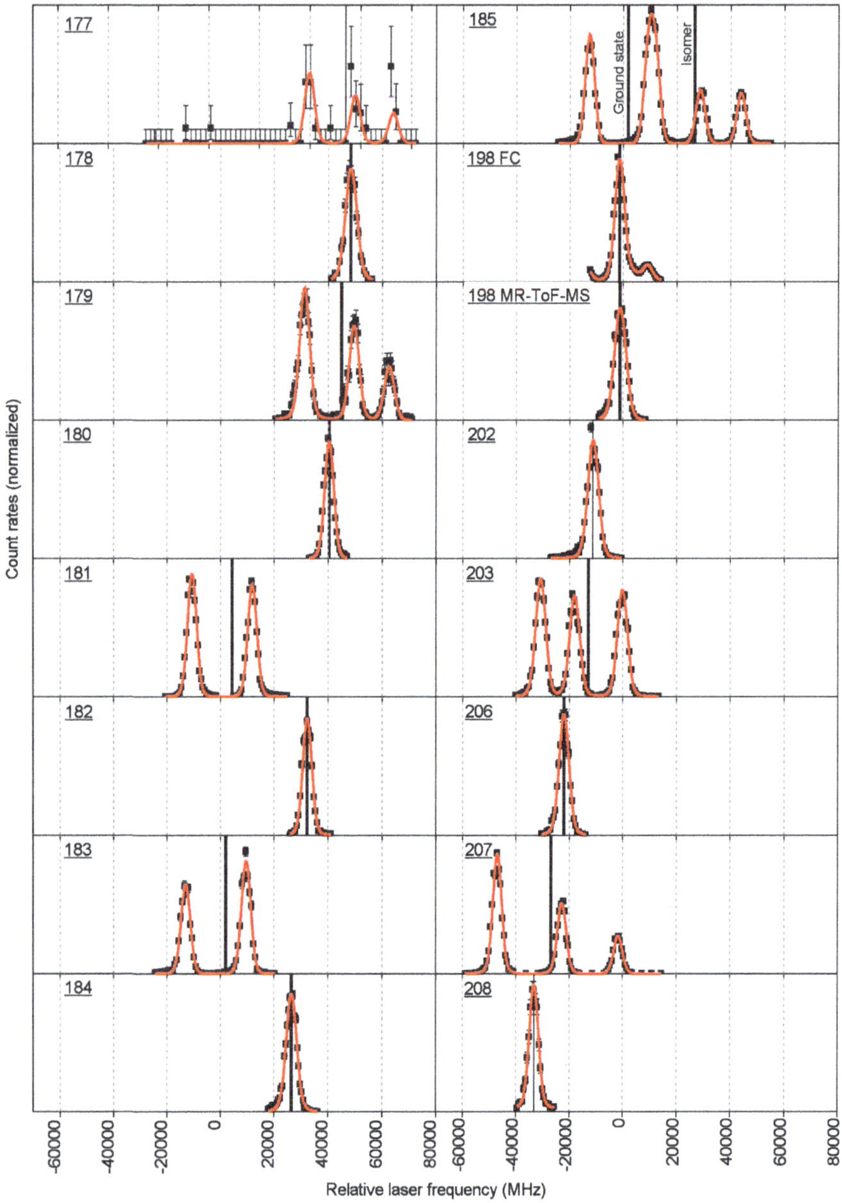

Fig. 8.2 Fitted laser scans of each of the mercury isotopes investigated in this work

Based on this analysis the isotope shifts, mean square charge radii and values for the hyperfine A and B factors were deduced. The final results are presented in Sect. 8.6. A selection of the scans are presented in Fig. 8.2.

8.5 Spin Assignments

The nuclear spin I^{π} could not be unambiguously determined from this work because of the number of HFS components (the result of the spectroscopic transition exciting between atomic levels with J = 0 and J = 1). A spin and parity of $I^{\pi} = 9/2+$ was used for the analysis of the measurements of ^{207}Hg, based on reported systematics [26] and the neutron shell $g_{9/2}$ being the first shell above the N=126 shell closure. The ground state of ^{179}Hg was assigned as $I^{\pi} = 7/2-$, based on the proposed decay scheme and the study of the α-decay of ^{183}Pb [27]. The ground state of ^{177}Hg was also taken to be $I^{\pi} = 7/2-$ based on the decay of the $I^{\pi} = 13/2+$ isomeric state [28]. The moments extracted from the HFS scans presented here support these spin assignments [29]. However, they are not discussed in this work and will form part of the thesis of S. Sels.

8.6 Summary of the Extracted Data

8.6.1 Isotope Shifts

The extracted $\delta\nu^{198,A}$ and $\delta\langle r^2\rangle^{198,A}$ from this work are presented in Table 8.2, together with the literature values for the previously measured isotopes taken from [23]. The $\delta\nu^{198,A}$ are compared graphically to literature values in Fig. 8.3. The uncertainties in the $\delta\nu^{198,A}$ of this work take into account the reproducibility of the CGs extracted from the ^{198}Hg reference measurements and the standard deviations of the CGs of each isotope. These uncertainties and additional weighting and uncertainty considerations for the CGs of ^{177}Hg and ^{208}Hg are discussed in the appendix of [25].

Figure 8.3 shows a reasonable agreement between the $\delta\nu^{198,A}$ from this work and the values reported by earlier studies and compiled in [23]. The significant uncertainty for the CG of ^{185}Hg from this work, can be at least partially attributed to the saturation of the detector of the MR-ToF MS hindering the deconvolution of ground state and isomer HFS components [25]. The discrepancy for ^{206}Hg is discussed in Sect. 10.3. The full $\delta\langle r^2\rangle^{198,A}$ plot is presented in Fig. 8.4.

Taking a global view of the mercury isotope chain as presented in Fig. 8.4, it can be seen that the results of this work are in reasonable agreement with the earlier measurements, including for ^{185}Hg and ^{206}Hg where the discrepancy in $\delta\nu^{198,A}$ was $\approx$500 MHz. The agreement with the results of previous experiments, obtained through the application of different experimental methods, provides a validation of the technique of in-source RIS in a VADIS anode cavity. Crucially, this validates the results for the newly measured isotopes.

Table 8.2 Compilation of the extracted isotope shifts from this work and those compiled by [23]. The data presented here is based on the fitting procedure outlined in Sect. 8.3. Further analysis was conducted within the collaboration, with the averaged results presented in [2]. The errors in the results presented here are purely experimental. To enable direct comparison, the literature values of $\delta\langle r^2\rangle^{198,A}$ are calculated from the $\delta\nu^{198,A}$ reported in [23], without scaling the electronic factor F_{254} for mass number A

A	$\delta\nu^{198,A}$ (MHz)	$\delta\nu^{198,A}$ (lit.) (MHz)	$\delta\langle r^2\rangle^{198,A}$ (fm^2)	$\delta\langle r^2\rangle^{198,A}$ (lit.) (fm^2)
177	55 100 (520)	–	–1.0674 (100)	–
178	49 400 (280)	–	–0.9584 (54)	–
179	46 170 (260)	–	–0.8958 (48)	–
180	41 350 (280)	–	–0.8026 (54)	–
181	5 370 (270)	5 560 (200)	–0.1093 (52)	–0.1130 (39)
182	33 450 (240)	33 250 (340)	–0.6497 (46)	–0.6458 (65)
183	3 150 (210)	3 310 (100)	–0.0658 (40)	–0.0689 (19)
184	27 690 (285)	27 720 (90)	–0.5382 (54)	–0.5386 (17)
185	3 210 (450)	3 710 (30)	–0.0662 (87)	–0.0759 (6)
185m	27 860 (350)	27 720 (110)	–0.5409 (67)	–0.5382 (21)
198	0	0	0	0
202	–10 050 (260)	–10 102.4 (4.2)	0.1948 (50)	0.1958 (1)
203	–11 830 (210)	–11 750 (180)	0.2294 (40)	0.2278 (35)
206	–20 890 (230)	–20 420 (80)	0.4047 (44)	0.3957 (15)
207	–25 710 (230)	–	0.4979 (44)	–
208	–31 970 (280)	–	0.6187 (54)	–

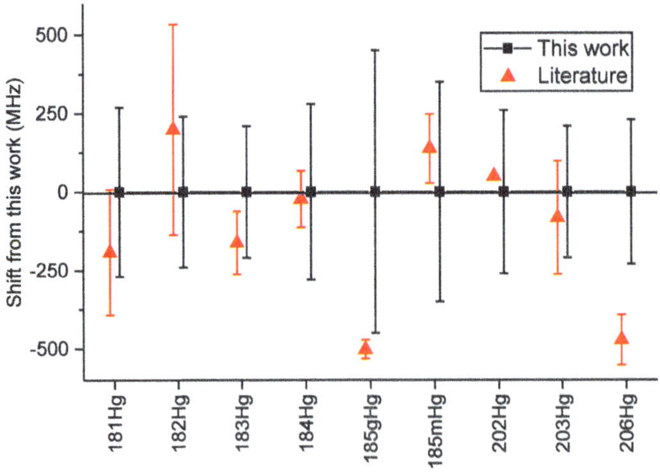

Fig. 8.3 A comparison of the extracted isotope shifts to literature values compiled in [23]. The "This work" and "Literature" data points have been offset to aid clarity

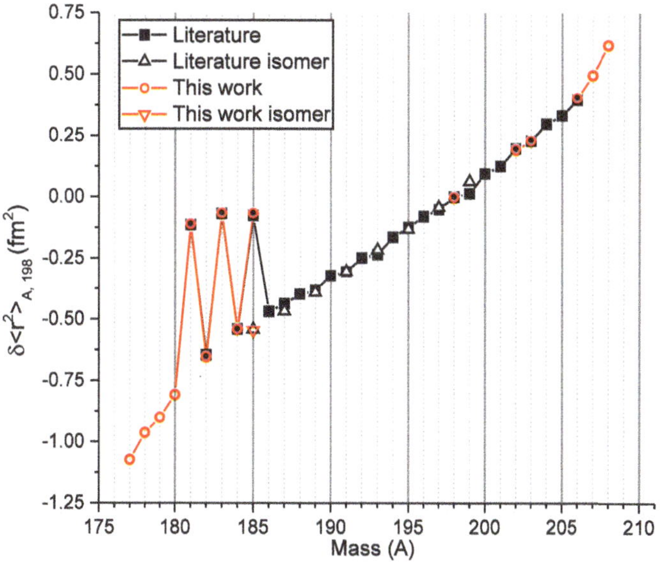

Fig. 8.4 Changes of mean square charge radii $\delta\langle r^2\rangle^{198,A}$ in the isotopic chain ^{177}Hg-^{208}Hg as a function of the mass number A and taking ^{198}Hg as a reference. The literature values were calculated from [23]. Spin assignments are discussed in Sect. 8.5

Table 8.3 Compilation of the hyperfine A and B factors from this work and compiled by [23]

A	Spin I^π	A factor (MHz)	A factor (lit) (MHz)	B factor (MHz)	B factor (lit) (MHz)
177	7/2−	−4 170 (160)	–	–	–
179	7/2−	−3 960 (50)	–	−530 (210)	–
181	1/2−	+15 030 (120)	+14 960 (250)	–	–
183	1/2−	+15 150 (80)	+15 380 (30)	–	–
185	1/2−	+14 620 (220)	+14 960 (70)	–	–
185m	13/2+	−2 290 (40)	−2 305 (19)	+310 (240)	−140 (230)
203	5/2−	−5 100 (40)	−4 991.33 (0.04)	+50 (60)	−249.2 (0.3)
207	9/2+	−4 510 (40)	–	+600 (160)	–

8.6.2 *Hyperfine A and B Factors*

The extracted hyperfine A and B factors from this work are presented in Table 8.3, with the literature values for the previously measured isotopes taken from [23]. The hyperfine A and B factors are compared graphically to literature values in Fig. 8.5.

The uncertainties for the hyperfine A and B factors from this work consider the standard deviation of the extracted values for each isotope and the fitting errors. The

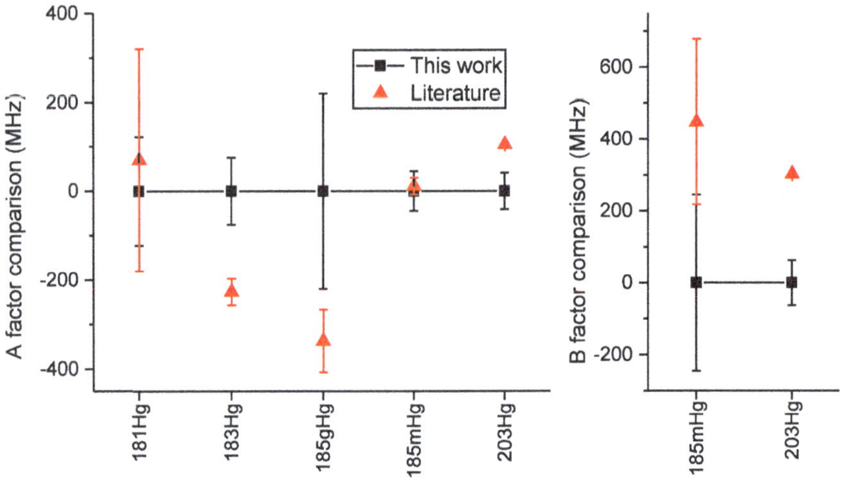

Fig. 8.5 Comparison of the extracted hyperfine A and B factors to the literature values compiled in [23]

extracted A and B factors from this work are in broad agreement with literature. However, there are discrepancies, particularly for the extracted A and B factors from ^{203}Hg. This indicates that there may be an additional contribution to the uncertainty that was not identified. As noted in Sect. 8.5, the investigation of the magnetic dipole and electric quadrupole moments extracted from this work will form part of the thesis of S. Sels and therefore will not be discussed further.

References

1. Gaffney LP, Day Goodacre T, Andreyev A, Seliverstov M, Althubiti N, Andel B, Antalic S, Atanasov D, Barzakh AE, Billowes J, Blaum K, Cocolios TE, Cubiss J, Farooq-Smith G, Fedosseev VN, Fedorov DV, Ferrer R, Flanagan KT, Ghys L, Gottberg A, Granados C, Herfurth F, Huyse M, Jenkins DG, Kisler D, Kreim S, Kron T, Kudryavtsev Yu, Lunney D, Lynch KM, Manea V, Marsh BA, Mendonca TM, Molkanov PL, Neidherr D, Raabe R, Raeder S, Ramos JP, Rapisarda E, Rosenbusch M, Rossel RE, Rothe S, Schweikhard L, Sels S, Stora T, Tsekhanovich I, Van Beveren C, Van Duppen P, Veinhard M, Wadsworth R, Wienholtz F, Welker A, Wendt K, Wilson GL, Witkins S, Wolf R, Zuber K (2014) In-source laser spectroscopy of mercury isotopes. CERN INTC 424. https://doi.org/10.17181/CERN.DUGM.L69R
2. Marsh BA, Day Goodacre T, Sels S, Tsunoda Y, Andel B, Andreyev AN, Althubiti NA, Atanasov D, Barzakh AE, Billowes J, Blaum K, Cocolios TE, Cubiss JG, Dobaczewski J, Farooq-Smith GJ, Fedorov DV, Fedosseev VN, Flanagan KT, Gaffney LP, Ghys L, Huyse M, Kreim S, Lunney D, Lynch KM, Manea V, Martinez Palenzuela Y, Molkanov PL, Otsuka T, Pastore A, Rosenbusch M, Rossel RE, Rothe S, Schweikhard L, Seliverstov MD, Spagnoletti P, Van Beveren C, Van Duppen P, Veinhard M, Verstraelen E, Welker A, Wendt K, Wienholtz F, Wolf RN, Zadvornaya A, Zuber K (2018) Characterization of the shape-staggering effect in mercury nuclei. Nat Phys. https://doi.org/10.1038/s41567-018-0292-8

3. Sels S, Day Goodacre T, Marsh BA, Pastore A, Ryssens W, Tsunoda Y, Althubiti N, Andel B, Andreyev AN, Atanasov D, Barzakh AE, Bender M, Billowes J, Blaum K, Cocolios TE, Cubiss JG, Dobaczewski J, Farooq-Smith GJ, Fedorov DV, Fedosseev VN, Flanagan KT, Gaffney LP, Ghys L, Heenen P-H, Huyse M, Kreim S, Lunney D, Lynch KM, Manea V, Martinez Palenzuela Y, Medonca TM, Molkanov PL, Otsuka T, Ramos JP, Rossel RE, Rothe S, Schweikhard L, Seliverstov MD, Spagnoletti P, Van Beveren C, Van Duppen P, Veinhard M, Verstraelen E, Welker A, Wendt K, Wienholtz F, Wolf RN, Zadvornaya A (2019) Shape staggering of midshell mercury isotopes from in-source laser spectroscopy compared with density-functional-theory and Monte Carlo shell-model calculations. Phys Rev C 99(4):044306. https://doi.org/10.1103/PhysRevC.99.044306
4. Day Goodacre T, Afanasjev AV, Barzakh AE, Marsh BA, Sels S, Ring P, Nakada H, Andreyev AN, Van Duppen P, Althubiti NA, Andel B, Atanasov D, Billowes J, Blaum K, Cocolios TE, Cubiss JG, Farooq-Smith GJ, Fedorov DV, Fedosseev VN, Flanagan KT, Gaffney LP, Ghys L, Huyse M, Kreim S, Lunney D, Lynch KM, Manea V, Martinez Palenzuela Y, Molkanov PL, Rosenbusch M, Rossel RE, Rothe S, Schweikhard L, Seliverstov MD, Spagnoletti P, Van Beveren C, Veinhard M, Verstraelen E, Welker A, Wendt K, Wienholtz F, Wolf RN, Zadvornaya A, Zuber K (2021) Laser spectroscopy of neutron-rich 207,208Hg isotopes: illuminating the kink and odd-even staggering in charge radii across the N=126 shell closure. Phys Rev Lett 126(3):032502. https://doi.org/10.1103/PhysRevLett.126.032502
5. Lettry J, Catherall R, Cyvoct G, Drumm P, Evensen AHM, Lindroos M, Jonsson OC, Kugler E, Obert J, Putaux JC, Sauvage J, Schindl K, Ravn H, Wildner E (1997) Release from ISOLDE molten metal targets under pulsed proton beam conditions. Nucl Instrum Methods Phys Res Sect B: Beam Interact Mater At 126(1–4):170–175. https://doi.org/10.1016/S0168-583X(96)01088-9
6. Lettry J, Catherall R, Drumm P, Van Duppen P, Evensen AHM, Focker GJ, Jokinen A, Jonsson OC, Kugler E, Ravn H (1997) Pulse shape of the ISOLDE radioactive ion beams. Nucl Instrum Methods Phys Res Sect B: Beam Interact Mater At 126(1–4):130–134. https://doi.org/10.1016/S0168-583X(96)01025-7
7. Lettry J, Arnau G, Benedikt M, Gilardoni S, Catherall R, Georg U, Cyvogt G, Fabich A, Jonsson O, Ravn H, Sgobba S, Bauer G, Brucherstseifer H, Graber T, Güdermann C, Ni L, Rastani R (2003) Effects of thermal shocks on the release of radioisotopes and on molten metal target vessels. Nucl Instrum Methods Phys Res Sect B: Beam Interact Mater At 204:251–256. https://doi.org/10.1016/S0168-583X(02)01919-5
8. Day Goodacre T, Billowes J, Catherall R, Cocolios TE, Crepieux B, Fedorov DV, Fedosseev VN, Gaffney LP, Giles T, Gottberg A, Lynch KM, Marsh BA, Mendonça TM, Ramos JP, Rossel RE, Rothe S, Sels S, Sotty C, Stora T, Van Beveren C, Veinhard M (2016) Blurring the boundaries between ion sources: the application of the RILIS inside a FEBIAD type ion source at ISOLDE. Nucl Instrum Methods Phys Res Sect B: Beam Interact Mater At 376:39–45. https://doi.org/10.1016/j.nimb.2016.03.005
9. Fedosseev V, Chrysalidis K, Day Goodacre T, Marsh B, Rothe S, Seiffert C, Wendt K (2017) Ion beam production and study of radioactive isotopes with the laser ion source at ISOLDE. J Phys G: Nucl Part Phys 44(8):084006. https://doi.org/10.1088/1361-6471/aa78e0
10. Penescu L, Catherall R, Lettry J, Stora T (2010) Development of high efficiency versatile arc discharge ion source at CERN ISOLDE. Rev Sci Instrum 81(2):02A906. https://doi.org/10.1063/1.3271245
11. Kirchner R, Roeckl E (1976) A novel ISOL ion source. Nucl Instrum Methods 139:291–296. https://doi.org/10.1016/0029-554X(76)90687-X
12. Kirchner R, Roeckl E (1976) Investigation of gaseous discharge ion sources for isotope separation on-line. Nucl Instrum Methods 133(2):187–204. https://doi.org/10.1016/0029-554X(76)90607-8
13. Sundell S, Ravn H (1992) Ion source with combined cathode and transfer line heating. Nucl Instrum Methods Phys Res Sect B: Beam Interact Mater At 70(1–4):160–164. https://doi.org/10.1016/0168-583X(92)95926-I

14. Fricke G, Heilig K (2005) 80-Hg mercury. Nuclear charge radii. Springer, Berlin, pp 1–9. https://doi.org/10.1007/10856314_82
15. Wolf RN, Beck D, Blaum K, Böhm Ch, Borgmann Ch, Breitenfeldt M, Herfurth F, Herlert A, Kowalska M, Kreim S, Lunney D, Naimi S, Neidherr D, Rosenbusch M, Schweikhard L, Stanja J, Wienholtz F, Zuber K (2012) On-line separation of short-lived nuclei by a multi-reflection time-of-flight device. Nucl Instrum Methods Phys Res Sect A: Accel Spectrom Detect Assoc Equip 686:82–90. https://doi.org/10.1016/j.nima.2012.05.067
16. Audi G, Kondev FG, Wang M, Huang WJ, Naimi S (2017) The NUBASE2016 evaluation of nuclear properties. Chin Phys C 41(3):030001. https://doi.org/10.1088/1674-1137/41/3/030001
17. De Witte H, Andreyev AN, Barré N, Bender M, Cocolios TE, Dean S, Fedorov D, Fedoseyev VN, Fraile LM, Franchoo S, Hellemans V, Heenen PH, Heyde K, Huber G, Huyse M, Jeppessen H, Köster U, Kunz P, Lesher SR, Marsh BA, Mukha I, Roussière B, Sauvage J, Seliverstov M, Stefanescu I, Tengborn E, Van de Vel K, Van de Walle J, Van Duppen P, Volkov Yu (2007) Nuclear charge radii of neutron-deficient lead isotopes beyond N=104 midshell investigated by in-source laser spectroscopy. Phys Rev Lett 98(11):112502. https://doi.org/10.1103/PhysRevLett.98.112502
18. OriginLab. 15.5.3 theory of nonlinear curve fitting. http://www.originlab.com/doc/Origin-Help/NLFit-Theory
19. Levenberg K (1944) A method for the solution of certain non-linear problems in least squares. Q Appl Math 2(2):164–168. http://www.jstor.org/stable/43633451
20. Marquardt DW (1963) An algorithm for least-squares estimation of nonlinear parameters. J Soc Ind Appl Math 11(2):431–441. https://doi.org/10.1137/0111030
21. Torbohm G, Fricke B, Rosén A (1985) State-dependent volume isotope shifts of low-lying states of group-II a and -II b elements. Phys Rev A 31(4):2038–2053. https://doi.org/10.1103/PhysRevA.31.2038
22. Seltzer EC (1969) K X-ray isotope shifts. Phys Rev 188(4):1916–1919. https://doi.org/10.1103/PhysRev.188.1916
23. Ulm G, Bhattacherjee SK, Dabkiewicz P, Huber G, Kluge H-J, Kühl T, Lochmann H, Otten EW, Wendt K, Ahmad SA, Klempt W, Neugart R (1986) Isotope shift of 182Hg and an update of nuclear moments and charge radii in the isotope range 181Hg-206Hg. Zeitschrift für Physik A Hadrons and nuclei 325(3):247–259. https://doi.org/10.1007/BF01294605
24. Fricke G, Heilig K (2005) Introduction. Nuclear charge radii. Springer, Berlin, pp 1–36. https://doi.org/10.1007/10856314_1
25. Day Goodacre T (2016) Developments of the ISOLDE RILIS for radioactive ion beam production and the results of their application in the study of exotic mercury isotopes. PhD thesis, The University of Manchester. https://doi.org/10.17181/CERN.NQQS.YS7S
26. Jonson B, Nielsen OB, Westgaard L, Zylicz J (1981) The decay of 207Hg to 207Tl. In: 4th international conference on nuclei far from stability V2, Helsingør, Denmark. CERN, pp 640–643. https://doi.org/10.5170/CERN-1981-009.640
27. Kondev FG, Carpenter MP, Janssens RVF, Lister CJ, Abu Saleem K, Ahmad I, Amro H, Caggiano J, Davids CN, Heinz A, Herskind B, Khoo TL, Lauritsen T, Ma WC, Ressler JJ, Reviol W, Riedinger LL, Sarantites DG, Seweryniak D, Siem S, Sonzogni AA, Varmette PG, Wiedenhöver I (2002) First observation of excited structures in neutron-deficient 179Hg: evidence for multiple shape coexistence. Phys Lett B 528(3–4):221–227. https://doi.org/10.1016/S0370-2693(02)01221-2
28. Melerangi A, Appelbe D, Page RD, Boardman HJ, Greenlees PT, Jones P, Joss DT, Julin R, Juutinen S, Kettunen H, Kuusiniemi P, Leino M, Muikku MH, Nieminen P, Pakarinen J, Rahkila P, Simpson J (2003) Shape isomerism and spectroscopy of 177Hg. Phys Rev C 68(4):041301. https://doi.org/10.1103/PhysRevC.68.041301
29. Barzakh AE (2016) Private communication. Technical report

Chapter 9
Charge Radii of Neutron Deficient Mercury

Abstract This chapter discusses the results of the isotope shift measurements of neutron deficient mercury, with respect to the chain of mercury isotopes, the region below Z = 82 and the predictions and interpretation of nuclear models.

9.1 A Cessation of Extreme Shape Staggering

A cessation of the extreme shape staggering in neutron deficient mercury can be clearly observed in the chain of isotope shifts between ^{177}Hg and ^{187}Hg presented in Fig. 9.1. These results represent the first definitive determination of the end point of the extreme shape staggering. The results presented in this chapter have been published in combination with theoretical calculations in [1, 2]. The trend of the charge radii of neutron deficient mercury isotopes is overlaid on lines corresponding to the quadrupole deformation parameter β_2 defined in Eq. 3.7, applying the revised coefficients of [3] for the calculation of the mean square radius of the spherical nuclei using the droplet model.

The strongly deformed ^{185,183,181}Hg nuclei correspond to $\beta_2 = 0.26 - 0.3$ and the weakly deformed nuclei to $\beta_2 = 0.13 - 0.17$. ^{181}Hg has the largest β_2 of the strongly deformed nuclei and ^{182}Hg the largest β_2 of the weakly deformed nuclei. The β_2 values decrease with decreasing N from A = 180, suggesting a decrease in the deformation of the weakly deformed isotopes.

The behaviour of the chain of mercury isotopes can additionally be considered in comparison to neighbouring elements, enabling a model-independent interpretation. This is presented in Fig. 9.2, where the isotope chains of thallium, mercury, gold and platinum (Z = 81 to 78 respectively) are overlaid on the charge radii of the closed proton shell (Z = 82) lead isotopes.

Taking a macroscopic view of the region around mercury, the charge radii shown in Fig. 9.2 could be understood as depicting the steadily decreasing influence of the Z = 82 shell closure. At Z = 81, the charge radii are near identical to that of the Z = 82 isotopes. Following the removal of one proton (Z = 80), the chain of isotopes

T. Day Goodacre, *Applied Laser Spectroscopy for Nuclear Physics*, Springer Theses,
https://doi.org/10.1007/978-3-030-73889-1_9

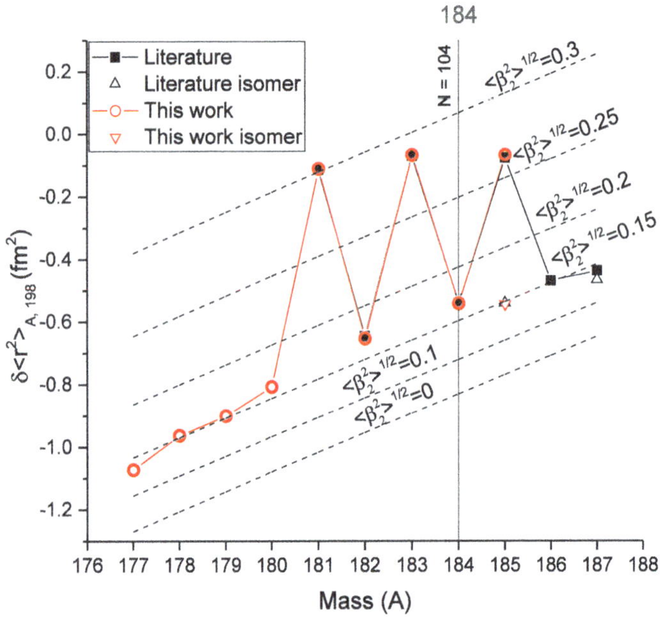

Fig. 9.1 Isotope shifts of neutron deficient mercury between ^{177}Hg and ^{187}Hg. The data is overlaid on lines of β_2^2 deformation, using the predictions of the droplet model [4] with updated parameters from [3] to calculate the predicted radii of spherical nuclei

exhibits extreme shape staggering between odd and even A isotopes, beginning at N = 106. In the isotope chain of Z = 79, there is a sudden isotope shift between N = 108 and N = 107. However, unlike its mercurial neighbour, the following even A isotopes remain deformed, resulting in an island of deformation. Finally, four protons from the Z = 82 shell closure (Z = 78), the isotopes steadily diverge from the trend of the lead charge radii, with a kink at N = 110.

Neutron shell effects can be inferred from the changes in charge radii with fixed Z. The increase in deformation with decreasing N following the N = 126 shell closure could be expected as the result of increasing numbers of valence neutrons. There is a reduction of this deformation, most notably now in the mercury isotopes, following the N = 104 midshell region. This could be interpreted as evidence in favour of a shell closure around N = 82, well beyond the proton drip line. It is interesting to note that the return to a weak deformation in the mercury nuclei at N = 100, is somewhat mirrored in the platinum isotopes, where a sharp decrease in deformation between N = 101 and N = 100 can be observed.

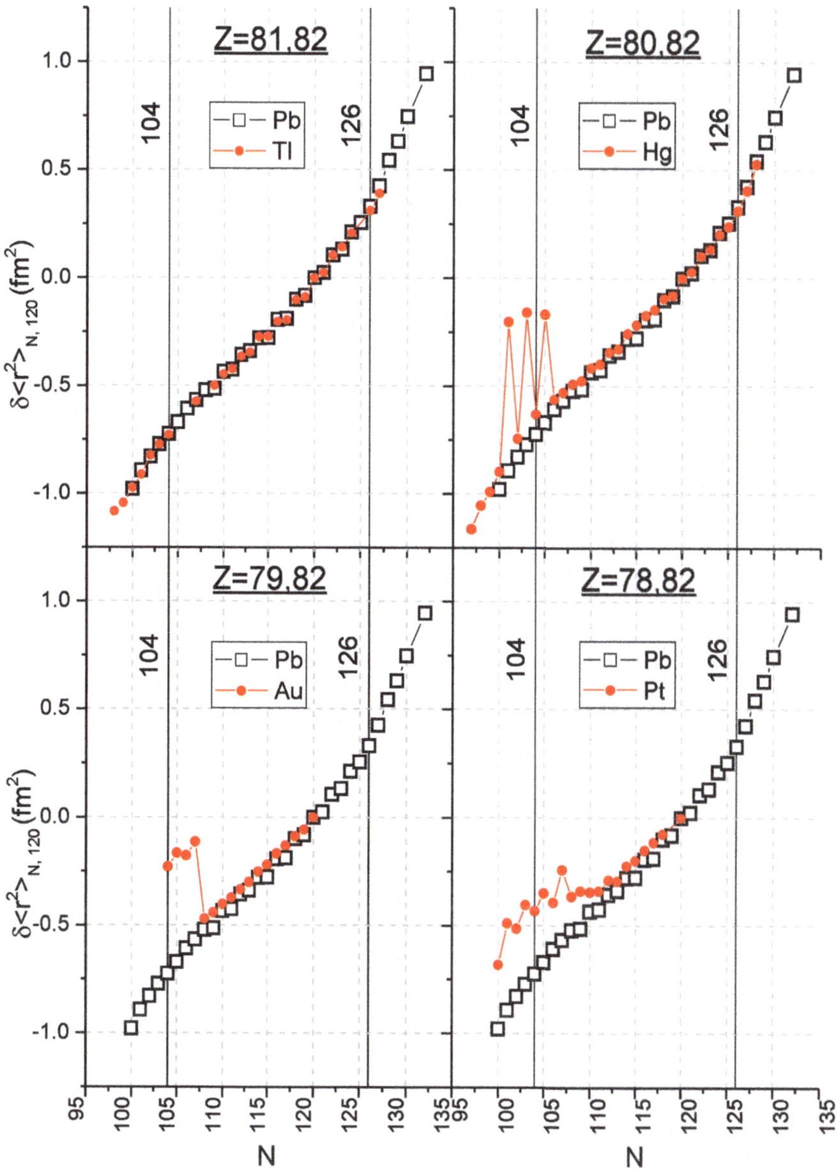

Fig. 9.2 Isotope shifts for the chains of isotopes from Z = 78 – 82. Data taken from [5–8] (Pb), [8–10] (Tl), [11] and this work (Hg), [8] (Au and Pt). The charge radii are all shown with respect to N = 120 for ease of comparison with the Au and Pt isotope chains

9.2 Discussion of the Results in the Framework of Non-optical Studies

As discussed in Sect. 1.6.2, experimental methods other than optical spectroscopy have yielded insights into the behaviour of the chain of mercury isotopes. Coulomb excitation experiments confirmed the coexistence of two structures at low energy for the light mass mercury isotopes [12]. The results of γ-spectroscopy experiments on the mercury isotope chain are shown in Fig. 9.3, reproduced from [13]. The level energies of the weakly deformed oblate band are shown in blue and the level energies of the strongly deformed prolate band are shown in red.

The intruder structure, understood to correspond to the deformed prolate deformation, is centred around N = 102. However, it is not completely symmetric, with the level energies of the intruder increasing more rapidly with decreasing N [14]. The level energies therefore indicate the disappearance of the coexisting prolate minima with decreasing N. However, it is not possible to determine from these results if the ground state shape staggering would cease at N = 100, or if there is sufficient symmetry around the N = 102 minima of the intruder level energies for the shape staggering to continue to N = 98. The suggestion of a weakly deformed ground state for the odd nuclei below A = 180, by [15–17], is in agreement with the results presented in this work.

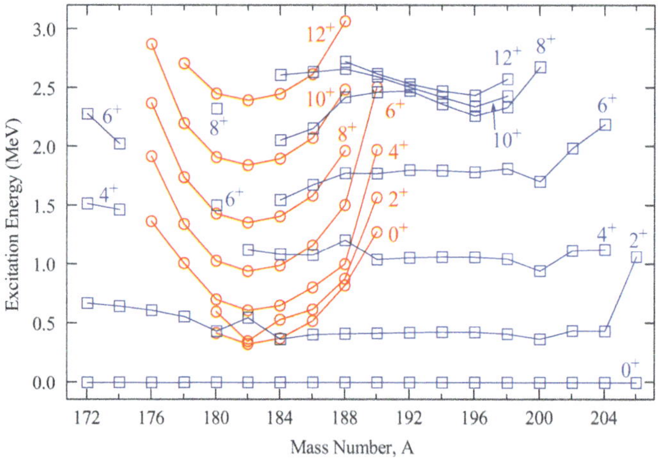

Fig. 9.3 Level energy systematics of even mercury isotopes between ^{172}Hg and ^{206}Hg. The level energies assigned to the oblate deformation are depicted as blue squares, those assigned to the intruder are depicted as red circles. Figure taken from [13]

9.3 Discussion of the Results in the Framework of Nuclear Models

Charge radii systematics enable the predictive powers of state of the art nuclear models to be validated, equally these models have an important role in the interpretation of experimental results. A number of nuclear models are capable of reproducing the competing shape configurations of the neutron deficient mercury nuclei. However, the reproduction of the pronounced odd-even shape staggering observed between ^{186}Hg and ^{180}Hg remains challenging. This can be understood as resulting from the complexity of the phenomena, where there are competing deformed shapes around the N = 104 mid shell, separated by a few hundred keV, with pairing effects sufficient to tip the balance between the two to determine the shape of the nuclear ground state. The charge radii measurements of neutron deficient mercury isotopes have been published in [1, 2] and compared with the results of Monte Carlo shell model (MCSM) [18], and Density Functional Theory (DFT) calculations.

9.3.1 *Comparison with the FRLDM*

In a wide ranging calculation, Möller et al. [19] used the FRLDM [20] (Sect. 2.2) to calculate the potential energy surfaces of 7206 nuclei between A = 31 and A = 290, with the results for 1224 even-even nuclei presented as potential energy contour plots of ϵ_2 Vs γ. Here ϵ_2 is the Nilsson quadrupole deformation parameter, which can be related to β_2 using an equation taken from [21], where

$$\beta_2 = \sqrt{\frac{\pi}{5}}\left(\frac{4}{3}\epsilon_2 + \frac{4}{9}\epsilon_2^2 + \frac{4}{27}\epsilon_2^3 + \frac{4}{81}\epsilon_2^4 + \ldots\right), \tag{9.1}$$

and γ describes the axial symmetry. $\gamma = 60$ corresponds to a purely oblate deformation, while $\gamma = 0$ corresponds to a purely prolate deformation. $0 < \gamma < 60$ corresponds to the degree of triaxial deformation. Contour plots taken from [19] for ^{176}Hg to ^{186}Hg are shown in Fig. 9.4. Minima deeper than 0.05 MeV are marked on each plot with black circles and saddle points with crosses. The numbers labelling the bold contour lines correspond potential energies with units of MeV, each contour line corresponds to an energy difference of 0.2 MeV.

The contour plots shown in Fig. 9.4 clearly depict competing oblate and prolate minima, with potential energies around 1 MeV. The prolate minima appears between N = 186 and N = 184 and disappears between N = 180 and N = 178. Were this to correspond exactly to the beginning and the end of the shape staggering, it would suggest the ground state of ^{179}Hg would have a strong prolate deformation. Applying Eq. 9.1, it can be seen that the deformations are in broad agreement with the droplet model deformation lines presented in Fig. 9.1. ^{185,183,181}Hg have $\beta_2 = 0.26 - 0.3$ ($\epsilon_2 = 0.23 - 0.26$) which corresponds to the strongly deformed prolate minima in

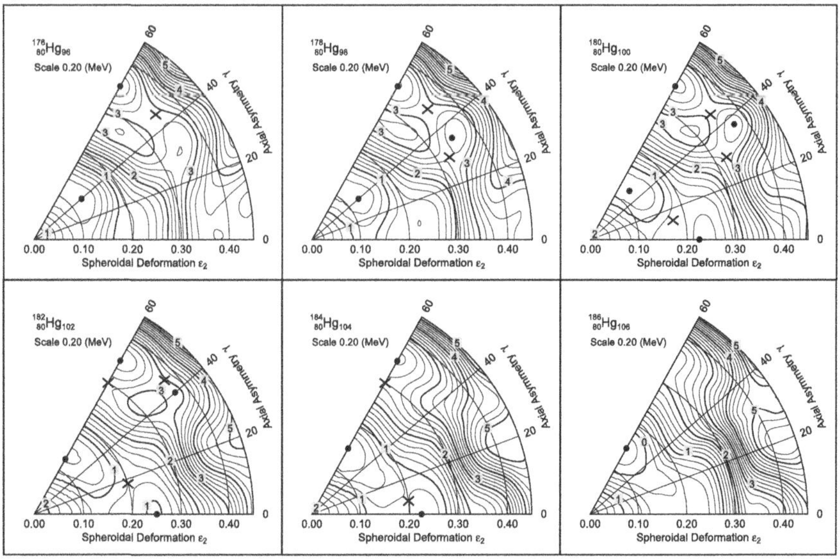

Fig. 9.4 FRLDM potential energy surface contour plots predicting the evolution of shape coexistence in even mercury isotopes from ^{176}Hg to ^{186}Hg. Parameters are defined in the text. Figure taken from [19]

the contour plots, the other isotopes have $\beta_2 = 0.13 - 0.17$ ($\epsilon_2 = 0.12 - 0.16$) corresponding to the weak oblate deformation in the FRLDM results. The contour plots enable the visualisation of the competing minima. However, the full results (included as a table in [19]) do not reproduce the odd-even shape staggering and indicate a weak oblate deformation of the nuclear ground state for both odd and even mercury isotopes around N = 104.

9.3.2 *Comparison with the IBM*

J.E. García-Ramos and K. Heyde applied the IBM (Sect. 2.4) to even A mercury isotopes [22]. The results were in good agreement with the energy level systematics shown in Fig. 9.3, similarly the model was able to well reproduce isotope shifts of mercury isotopes, though only even isotopes between N = 104 and N = 114. Predictions were not reported beyond N = 104. The calculations indicated two competing shapes corresponding to a weak oblate deformation and a strong prolate deformation. However, the IBM results also indicated that the ground state of the even isotopes had a pronounced prolate deformation between $100 \leq N \leq 104$, with the energy minima degenerate at N = 106. This was in contradiction to their calculations for the isotope shift of ^{184}Hg. The IBM is heavily influenced by the experimental data used in the formulation of the model [23], a particularly good example being the determination of the lowest lying level energy of the intruder band observed in neutron deficient lead [24].

9.3.3 *Comparison with (Relativistic) Mean Field Calculations*

The mercury isotopes have been investigated with both mean field and relativistic mean field calculations (Sect. 2.5). Mean field calculations by Yao et al. [25] and Nomura et al. [26] were able to reproduce the competing oblate and prolate shapes. However, the results corresponded to a strongly deformed prolate ground state for the even isotopes between $100 \leq N \leq 106$ and $98 \leq N \leq 106$ respectively. In relativistic mean field calculations, Nikšić et al. [27] were able to reproduce the weakly oblate ground states for even mass neutron deficient mercury isotopes. However, this was only possible using an effective interaction parametrization tailored specifically to the $Z = 78 - 82$ region. Therefore the authors did not consider it to be a general effective relativistic mean-field interaction. The calculations were not extended to the odd mercury isotopes. A more recent and wide ranging set of calculations by Agbemava et al. [28] determined the even isotopes to have a strong prolate deformation for $100 \leq N \leq 108$.

Following the in-source spectroscopy results reported here, axial Hartree-Fock-Bogoliubov calculations were undertaken for both odd and even mercury isotopes [2]. Through a careful tuning of the parameters, their calculations were capable of reproducing the pronounced shape staggering behaviour.

9.4 Conclusion

The mean square charge radii measurements were extended four isotopes further into the neutron deficient region of the nuclear chart, down to ^{177}Hg. These measurements provided a definitive demonstration of the cessation of the extreme odd-even staggering in the neutron deficient mercury isotopes. This reduction in deformation could be interpreted as evidence for the robustness of a shell closure around $N = 82$, beyond the proton drip line.

Nuclear models are capable of demonstrating the competing oblate and prolate deformations, providing a valuable tool for the interpretation of the results. Nuclear models are also capable of reproducing the lowering of the potential energy of the well deformed prolate minima around the $N = 104$ midshell. Recent Hartree-Fock-Bogoliubov and MCSM calculations, based on the results reported here, were able to reproduce the extreme shape staggering behaviour in the neutron deficient mercury isotopes.

References

1. Marsh BA, Day Goodacre T, Sels S, Tsunoda Y, Andel B, Andreyev AN, Althubiti NA, Atanasov D, Barzakh AE, Billowes J, Blaum K, Cocolios TE, Cubiss JG, Dobaczewski J, Farooq-Smith GJ, Fedorov DV, Fedosseev VN, Flanagan KT, Gaffney LP, Ghys L, Huyse M, Kreim S, Lunney D, Lynch KM, Manea V, Martinez Palenzuela Y, Molkanov PL, Otsuka T, Pastore A, Rosenbusch M, Rossel RE, Rothe S, Schweikhard L, Seliverstov MD, Spagnoletti P, Van Beveren C, Van Duppen P, Veinhard M, Verstraelen E, Welker A, Wendt K, Wienholtz F, Wolf RN, Zadvornaya A, Zuber K (2018) Characterization of the shape-staggering effect in mercury nuclei. Nat. Phys. https://doi.org/10.1038/s41567-018-0292-8
2. Sels S, Day Goodacre v, Marsh BA, Pastore A, Ryssens W, Tsunoda Y, Althubiti N, Andel B, Andreyev AN, Atanasov D, Barzakh AE, Bender M, Billowes J, Blaum K, Cocolios TE, Cubiss JG, Dobaczewski J, Farooq-Smith GJ, Fedorov DV, Fedosseev VN, Flanagan KT, Gaffney LP, Ghys L, Heenen P-H, Huyse M, Kreim S, Lunney D, Lynch KM, Manea V, Martinez Palenzuela Y, Medonca TM, Molkanov PL, Otsuka T, Ramos JP, Rossel RE, Rothe S, Schweikhard L, Seliverstov MD, Spagnoletti P, Van Beveren C, Van Duppen P, Veinhard M, Verstraelen E, Welker A, Wendt K, Wienholtz F, Wolf RN, Zadvornaya A (2019) Shape staggering of midshell mercury isotopes from in-source laser spectroscopy compared with density-functional-theory and Monte Carlo shell-model calculations. Phys Rev C 99(4):044306. https://doi.org/10.1103/PhysRevC.99.044306
3. Berdichevsky D, Tondeur F (1985) Nuclear core densities, isotope shifts, and the parametrization of the droplet model. Zeitschrift für Physik A Hadrons and nuclei 322(1):141–147. https://doi.org/10.1007/BF01412027
4. Myers WD, Schmidt KH (1983) An update on droplet-model charge distributions. Nuclear Phys A 410(1):61–73. https://doi.org/10.1016/0375-9474(83)90401-3
5. Anselment M, Faubel W, Göring S, Hanser A, Meisel G, Rebel H, Schatz G (1986) The odd-even staggering of the nuclear charge radii of Pb isotopes. Nuclear Phys A 451(3):471–480. https://doi.org/10.1016/0375-9474(86)90071-0
6. Dutta SB, Kirchner R, Klepper O, Kühl T, Marx D, Sprouse GD, Menges R, Dinger U, Huber G, Schröder S (1991) Measurement of isotope shift and hyperfine splitting of 190, 191, 193, 197Pb isotopes by collinear laser spectroscopy. Zeitschrift für Physik A Hadrons and nuclei 341(1):39–45. https://doi.org/10.1007/BF01281272
7. De Witte H, Andreyev AN, Barré N, Bender M, Cocolios TE, Dean S, Fedorov D, Fedoseyev VN, Fraile LM, Franchoo S, Hellemans V, Heenen PH, Heyde K, Huber G, Huyse M, Jeppessen H, Köster U, Kunz P, Lesher SR, Marsh BA, Mukha I, Roussière B, Sauvage J, Seliverstov M, Stefanescu I, Tengborn E, Van de Vel K, Van de Walle J, Van Duppen P, Volkov Yu (2007) Nuclear charge radii of neutron-deficient lead isotopes beyond N = 104 midshell investigated by in-source laser spectroscopy. Phys Rev Lett 98(11):112502. https://doi.org/10.1103/PhysRevLett.98.112502
8. Angeli I, Marinova KP (2013) Table of experimental nuclear ground state charge radii: an update. Atomic Data Nuclear Data Tables 99(1):69–95. https://doi.org/10.1016/j.adt.2011.12.006
9. Barzakh AE, Batist LKh, Fedorov DV, Ivanov VS, Mezilev KA, Molkanov PL, Moroz FV, Orlov SYu, Panteleev VN, Volkov YuM (2013) Changes in the mean-square charge radii and magnetic moments of neutron-deficient Tl isotopes. Phys Rev C 88(2):024315. https://doi.org/10.1103/PhysRevC.88.024315
10. Van Beveren C (2016) Laser-assisted decay and optical spectroscopy studies of neutron-deficient thallium isotopes. PhD thesis, KU Leuven. https://lirias.kuleuven.be/handle/123456789/529374
11. Ulm G, Bhattacherjee SK, Dabkiewicz P, Huber G, Kluge H-J, Kühl T, Lochmann H, Otten EW, Wendt K, Ahmad SA, Klempt W, Neugart R (1986) Isotope shift of 182Hg and an update of nuclear moments and charge radii in the isotope range 181Hg–206Hg. Zeitschrift für Physik A Hadrons Nuclei 325(3):247–259. https://doi.org/10.1007/BF01294605

12. Bree N, Wrzosek-Lipska K, Petts A, Andreyev A, Bastin B, Bender M, Blazhev A, Bruyneel B, Butler PA, Butterworth J, Carpenter MP, Cederkäll J, Clément E, Cocolios TE, Deacon A, Diriken J, Ekström A, Fitzpatrick C, Fraile LM, Fransen Ch, Freeman SJ, Gaffney LP, García-Ramos JE, Geibel K, Gernhäuser R, Grahn T, Guttormsen M, Hadinia B, Hadyńska-Klęk K, Hass M, Heenen P-H, Herzberg R-D, Hess H, Heyde K, Huyse M, Ivanov O, Jenkins DG, Julin R, Kesteloot N, Kröll Th, Krücken R, Larsen AC, Lutter R, Marley P, Napiorkowski PJ, Orlandi R, Page RD, Pakarinen J, Patronis N, Peura PJ, Piselli E, Rahkila P, Rapisarda E, Reiter P, Robinson AP, Scheck M, Siem S, Singh Chakkal K, Smith JF, Srebrny J, Stefanescu I, Tveten GM, Van Duppen P, Van de Walle J, Voulot D, Warr N, Wenander F, Wiens A, Wood JL, Zielińska M (2014) Shape coexistence in the neutron-deficient even-even 182-188Hg isotopes studied via Coulomb excitation. Phys Rev Lett 112(16):162701. https://doi.org/10.1103/PhysRevLett.112.162701
13. Gaffney LP, Hackstein M, Page RD, Grahn T, Scheck M, Butler PA, Bertone PF, Bree N, Carroll RJ, Carpenter MP, Chiara CJ, Dewald A, Filmer F, Fransen C, Huyse M, Janssens RVF, Joss DT, Julin R, Kondev FG, Nieminen P, Pakarinen J, Rigby SV, Rother W, Van Duppen P, Watkins HV, Wrzosek-Lipska K, Zhu S (2014) Shape coexistence in neutron-deficient Hg isotopes studied via lifetime measurements in 184,186Hg and two-state mixing calculation. Phys Rev C 89(2):024307. https://doi.org/10.1103/PhysRevC.89.024307
14. Julin R, Helariutta K, Muikku M (2001) Intruder states in very neutron-deficient Hg, Pb and Po nuclei. J Phys G: Nuclear Part Phys 27(7):R109–R139. https://doi.org/10.1088/0954-3899/27/7/201
15. Kondev FG, Carpenter MP, Janssens RVF, Lister CJ, Abu Saleem K, Ahmad I, Amro H, Caggiano J, Davids CN, Heinz A, Herskind B, Khoo TL, Lauritsen T, Ma WC, Ressler JJ, Reviol W, Riedinger LL, Sarantites DG, Seweryniak D, Siem S, Sonzogni AA, Varmette PG, Wiedenhöver I (2002) First observation of excited structures in neutron-deficient 179Hg: evidence for multiple shape coexistence. Phys Lett B 528(3–4):221–227. https://doi.org/10.1016/S0370-2693(02)01221-2
16. Melerangi A, Appelbe D, Page RD, Boardman HJ, Greenlees PT, Jones P, Joss DT, Julin R, Juutinen S, Kettunen H, Kuusiniemi P, Leino M, Muikku MH, Nieminen P, Pakarinen J, Rahkila P, Simpson J (2003) Shape isomerism and spectroscopy of 177Hg. Phys Rev C 68(4):041301. https://doi.org/10.1103/PhysRevC.68.041301
17. Jenkins DG, Andreyev AN, Page RD, Carpenter MP, Janssens RVF, Lister CJ, Kondev FG, Enqvist T, Greenlees PT, Jones PM, Julin R, Juutinen S, Kettunen H, Kuusiniemi P, Leino M, Leppännen A-P, Nieminen P, Pakarinen J, Rahkila P, Uusitalo J, O'Leary CD, Raddon P, Simons A, Wadsworth R, Joss DT (2002) Confirmation of triple shape coexistence in 179Hg. Phys Rev C 66(1):011301. https://doi.org/10.1103/PhysRevC.66.011301
18. Otsuka T, Honma M, Mizusaki T, Shimizu N, Utsuno Y (2001) Monte Carlo shell model for atomic nuclei. Prog. Part. Nucl. Phys. 47(1):319–400. https://doi.org/10.1016/S0146-6410(01)00157-0
19. Möller P, Sierk AJ, Bengtsson R, Sagawa H, Ichikawa T (2012) Nuclear shape isomers. Atomic Data Nucl Data Tables 98(2):149–300. https://doi.org/10.1016/j.adt.2010.09.002
20. Moller P, Nix JR, Myers WD, Swiatecki WJ (1995) Nuclear ground-state masses and deformations. Atomic Data Nucl Data Tables 59(2):185–381. https://doi.org/10.1006/adnd.1995.1002
21. Firestone R (1999) Table of Isotopes (1999 update with CD-ROM), 8th edn. Wiley, Hoboken. http://eu.wiley.com/WileyCDA/WileyTitle/productCd-0471356336.html
22. García-Ramos JE, Heyde K (2014) Nuclear shape coexistence: a study of the even-even Hg isotopes using the interacting boson model with configuration mixing. Phys Rev C 89(1):014306. https://doi.org/10.1103/PhysRevC.89.014306
23. Wrzosek-Lipska K, Gaffney LP (2016) Unique and complementary information on shape coexistence in the neutron-deficient Pb region derived from Coulomb excitation. J Phys G: Nucl Part Phys 43(2):024012. https://doi.org/10.1088/0954-3899/43/2/024012
24. Barfield AF, Barrett BR (1984) The interacting boson model and the structure of 182Hg. Phys Lett B 149(4–5):277–278. https://doi.org/10.1016/0370-2693(84)90405-2

25. Yao JM, Bender M, Heenen PH (2013) Systematics of low-lying states of even-even nuclei in the neutron-deficient lead region from a beyond-mean-field calculation. Phys Rev C 87(3):034322. https://doi.org/10.1103/PhysRevC.87.034322
26. Nomura K, Rodríguez-Guzmán R, Robledo LM (2013) Shape evolution and the role of intruder configurations in Hg isotopes within the interacting boson model based on a Gogny energy density functional. Phys Rev C 87(6):064313. https://doi.org/10.1103/PhysRevC.87.064313
27. Nikšić T, Vretenar D, Ring P, Lalazissis GA (2002) Shape coexistence in the relativistic Hartree–Bogoliubov approach. Phys Rev C 65(5):054320. https://doi.org/10.1103/PhysRevC.65.054320
28. Agbemava SE, Afanasjev AV, Ray D, Ring P (2014) Global performance of covariant energy density functionals: ground state observables of even-even nuclei and the estimate of theoretical uncertainties. Phys Rev C 89(5):054320. https://doi.org/10.1103/PhysRevC.89.054320

Chapter 10
Charge Radii of Neutron Rich Mercury

Abstract This chapter discusses the results of the isotope shift measurements of neutron rich mercury across the N = 126 shell closure. The characteristic kink in the mean square charge radii centred at N = 126 is characterised, and compared to the one present in the chain of lead isotopes.

10.1 Introduction

The crossing of the N = 28, 50, 82 and 126 shell closures correspond to an increase in the gradient of the isotope shifts (a "kink") with increasing N [1]. The reproduction of the kink is often used as a benchmark for nuclear models [2]. Sharma et al. [3] were the first to reproduce the kink, using relativistic mean field calculations. The applicability of relativistic mean field models to reproducing the kink is understood to be related to the automatic inclusion of the spin-orbit interaction [4]. The kink is the result of the attraction between protons and neutrons. The spin-orbit interaction affects the single particle neutron levels, the occupancy of which affects the proton wave function and thus the mean square charge radii. Optical spectroscopy data on the mean square charge radii of even lead and polonium isotopes, was used to investigate the gradient of the kink at the crossing of the N = 126 shell closure by Goddard et al. [5]. Their calculations determined that the occupation of the $1i_{11/2}$ neutron orbital was required to reproduce the gradient of the kink in the lead and polonium isotope chains.

This work, which extends the mercury isotope shift measurements further into the neutron rich region of the nuclear chart, reveals the characteristic kink and enables the determination of the change in the gradient of the isotope shifts. Mercury is only the third isotope chain for which both N = 127 and N = 128 have been measured after lead and bismuth [6] and the first below the Z = 82 proton shell closure. The current capability of nuclear models in this region of the nuclear chart is often limited to even-even isotopes. Therefore, the determination of the isotope shifts of the even mercury isotopes is particularly important for the benchmarking of such models. The results reported here are presented in Fig. 10.1 where the kink in the trend of the

T. Day Goodacre, *Applied Laser Spectroscopy for Nuclear Physics*, Springer Theses,
https://doi.org/10.1007/978-3-030-73889-1_10

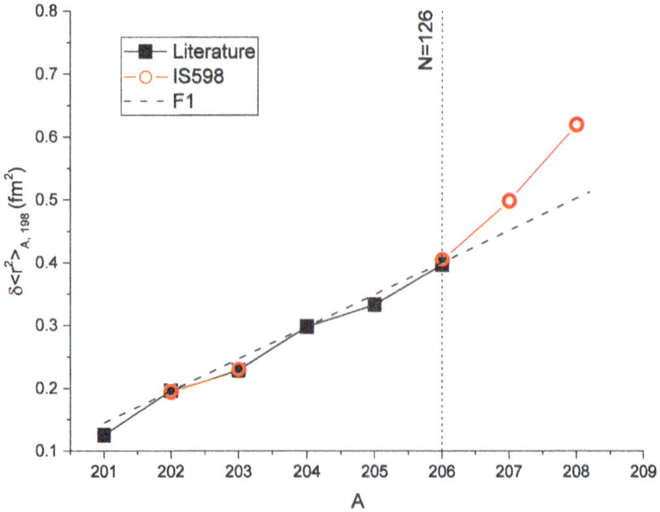

Fig. 10.1 Isotope shifts of stable and neutron rich mercury isotopes ^{201}Hg-^{208}Hg. Data taken from this work (hollow red circles) and calculated from the $\delta\nu^{198,A}$ reported in [7] (solid black squares)

isotope shifts is clearly visible, a line is drawn through the even isotopes $\leqslant$ N = 126 to guide the eye.

Methods for defining the kink are presented in Sect. 10.2. The discrepancy between the $\delta\langle r^2\rangle^{198,206}$ reported in this work and the previously reported values is investigated in Sect. 10.3. Finally, the gradient of the kink in the trend of the charge radii of mercury isotopes at N = 126 is determined in Sect. 10.4 and compared to the predictions of nuclear models. The results presented in this chapter have been published in combination with theoretical calculations in [8].

10.2 Defining the Kink

Before investigating the magnitude of the kink in the charge radii systematics, it is necessary to find an objective definition of this phenomenon. There are a number of possible approaches:

- The change in gradient considering even-N isotopes adjacent to the kink $R(m_{(2N)})$ where

$$R(m_{(2N)}) = \frac{\delta\langle r^2\rangle^{N=126,124}}{\delta\langle r^2\rangle^{N=128,126}}. \tag{10.1}$$

  This approach offers the best determination of the kink, as no odd-even staggering effects are included and the gradient is not affected by increasing deformation moving away from the neutron shell closure. However, for $84 \leq Z \geq 88$ $t_{1/2}$ at $N = 128$ is $< 300\,\mu s$, hindering potential laser spectroscopy measurements.

- Determination of the change in gradient considering isotopes adjacent to the kink $R(m_{(1N)})$ where

$$R(m_{(1N)}) = \frac{\delta \langle r^2 \rangle^{N=126,125}}{\delta \langle r^2 \rangle^{N=127,126}}. \tag{10.2}$$

  Considering the adjacent odd-N isotopes expands the range of elements for which laser spectroscopy is feasible using conventional methods at thick-target ISOL facilities: for $80 \leq Z \geq 88$ $t_{1/2}$ at $N = 127$ is > 1.5 ms. However, considering the adjacent odd-N isotopes includes a contribution from the odd-even staggering, which makes a direct comparison between isotope chains more challenging. The contribution of the odd-even staggering is additionally an issue when determining the kink in the trend of isotopes where there is, or may be, a reversal of the odd-even staggering beyond the neutron shell closure, as has been reported for $Z = 86$, 88 [9].
- A final option is a linear fit of the even-N isotopes before and after the kink $R(m_{(fit)})$. This method has the advantage of avoiding the influence of odd-even staggering. Additionally, even-N isotopes with longer half lives can be used to determine the gradient, enabling a gradient to be determined for $80 \leq Z \leq 88$. However, changes in deformation moving away from the shell closure can result in a non-linear trend of isotope shifts, reducing the accuracy of the determination of the kink. Additionally, the gradient of the trend of isotope shifts is not consistent immediately prior to and following the $N = 126$ shell closure. Considering the lead isotope chain, the gradient increases between $N = 124$ and $N = 126$ compared to the even-N isotopes between $N = 120$ and $N = 124$, the same behaviour is seen following the shell closure, with an increased gradient between and $N = 126$ and $N = 128$ compared to the even-N isotopes between $N = 128$ and $N = 132$. This is illustrated in Fig. 10.3 for the lead isotope chain, with the gradients determined from the various fitting options presented in Table 10.1.

From Fig. 10.2 and Table 10.1, it is evident that the different approaches for defining the kink yield significantly differing values. In Fig. 10.2 it can be seen that linear fits of the even isotopes $N = 120-124$ and $N = 128-132$ (shown in red) do not meet at $N = 126$, therefore failing to provide a value for the kink at the shell closure. From Table 10.1 it can be seen that the $R(m_{(fit)})$ method overestimated the kink defined as $R(m_{(2N)})$ by 6.5 (7)%. The effects of the odd-even staggering when considering $R(m_{(1N)})$ (shown in blue) are significantly more pronounced, using the values from Table 10.1 this method underestimates the kink defined as $R(m_{(2N)})$

Table 10.1 A comparison of kink determination methods for the lead isotope chain, data from [1]

Element	$R(m_{(2N)})$	$R(m_{(fit)}) = \frac{\delta \langle r^2 \rangle^{N=120,124}}{\delta \langle r^2 \rangle^{N=128,132}}$	$R(m_{(1N)})$
Pb	1.787 (6)	1.904 (11)	1.272 (8)

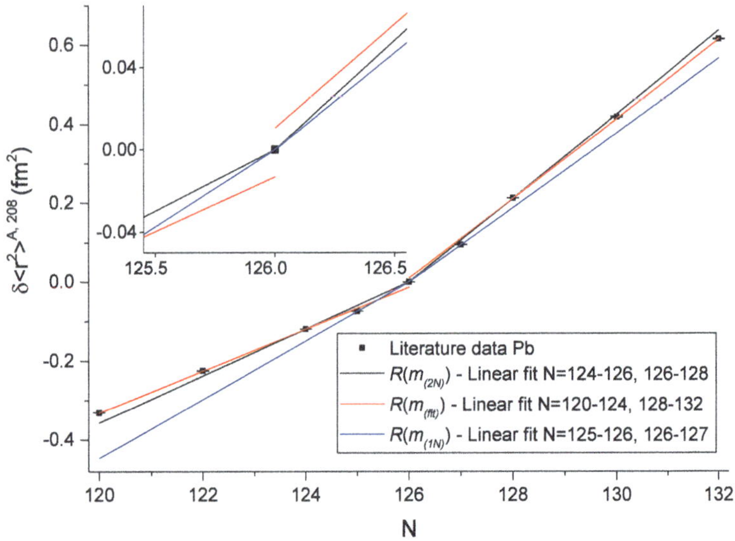

Fig. 10.2 Linear fits of the mean square charge radii at the crossing of the N = 126 shell closure in the lead isotope chain (data from [1]). $R(m_{(2N)})$ is the change in gradient considering N = 126 and the adjacent even-N isotopes (N = 124, 126 and N = 126, 128), $R(m_{(fit)})$ is the change in gradient considering the even-N isotopes before and after N = 126 (N = 120, 122, 124 and N = 128, 130, 132) and $R(m_{(1N)})$ is the change in gradient considering N = 126 and the immediately adjacent isotopes (N = 125, 126 and N = 126, 127)

by 29.1 (7)%. The $R(m_{(1N)})$ results require a multiplication factor to accurately reproduce the kink. However, the magnitude of the odd-even staggering varies along chains of isotopes and across shell closures, consequently any accurate correction would require knowledge of isotopes N = 124, 126, 128 which negates the usefulness of the correction.

The mercury isotope chain is therefore a particularly good case to investigate the kink at the N = 126 shell closure as it is possible to determine $R(m_{(2N)})$: the optimal method by which to benchmark nuclear models seeking to reproduce the trend in mean square charge radii across the shell closure.

10.3 ^{206}Hg

The most important isotope in the determination of the kink is N = 126 (^{206}Hg in the mercury isotope chain). Therefore the discrepancy between the $\delta\nu^{198,A}$ determined in this work and previous measurements (Fig. 8.3) is of particular concern. The literature value $\delta\nu^{198,206} = -20\,420\,(80)$ MHz, included in Table 8.2, was reported in the thesis of Dabkiewicz [10] and included in the 1986 mercury charge radii compilation of Ulm et al. [7]. However, the thesis value was not included in the 2005 compendium of

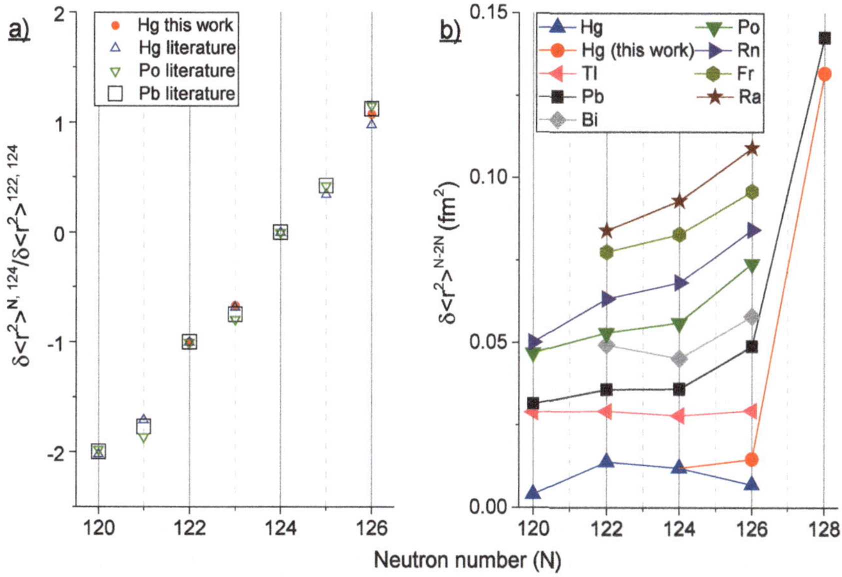

Fig. 10.3 **a** Relative variation in the mean square charge radii, comparing Hg (data from this work and calculated from the $\delta\nu^{198,A}$ reported in [7]), Pb (data taken from [1]), Po (data taken from [1]) between N = 120 and N = 128. **b** The difference in mean square charge radii between even-N isotopes for Z = 80 − 84, 86 − 88. The isotope chains are artificially displaced to aid interpretation. The mercury data is taken from this work and calculated from the $\delta\nu^{198,A}$ reported in [7], all other data is taken from [1]

charge radii by Fricke and Heilig [11], they instead used a previously published value of $\delta\nu^{204,206} = -4\ 700\ (500)\,\text{MHz} \approx \delta\nu^{198,206} = -20\ 010\ (500)\,\text{MHz}$ from [12]. These shifts are respectively 470 MHz and 880 MHz smaller than the $\approx \delta\nu^{198,206} = -20\ 890\ (230)\,\text{MHz}$ reported in this work.

There were multiple mercury isotopes present in the A = 206 mass selected ion beam during the previous measurements of ^{206}Hg [10, 12]. If the tail of the dominant ^{204}Hg signal was not fully accounted for in the fitting of the ^{206}Hg resonance, the effect would be a shift in the CG towards a smaller $\delta\nu^{198,A}$, and thus a smaller $\delta\langle r^2\rangle^{198,A}$. Such an issue would have the effect of shifting the value obtained in the same direction as that of the discrepancy described here.

Considering the systematics of the region, there is a reasonable agreement between the relative mean square charge radii of the lead (Z = 82) and polonium (Z = 84) isotopes approaching the N = 126 shell closure. The relative mean square charge radii of mercury (Z = 80) isotopes from this work and the values reported in [7] are compared to the lead and polonium isotopes in Fig. 10.3a. Additionally, the difference in mean square charge radii of even-N isotopes of Z = 80 − 84, 85 − 88 is compared in Fig. 10.3b.

In Fig. 10.3a, it can be seen that there is increasingly good agreement for the odd-A and even-A isotopes approaching N = 126 (isotopes of N = 122, 124 are overlapped by the nature of the comparison). The value of $\delta\langle r^2\rangle^{A,126}$ for mercury obtained in this work is in better agreement with the systematics of the region than the previous value reported in [7]. From Fig. 10.3b, it can be seen that the difference in the mean square charge radii increases between N = 124 and N = 126 in the isotope chains of Z = 81 – 84, 86 – 88. Prior to this work, only the mercury isotope chain exhibited a decrease. If we replace the earlier value for ^{206}Hg with the new value from this work, the mercury isotope chain is in agreement with the systematics of the region.

These discussions are not conclusive. However, since there is not an identified cause to doubt our measurements, which are in agreement with other literature data, it is deemed appropriate to use the value determined in this work in the following determination of the kink.

10.4 Crossing the N = 126 Shell Closure

The kink at the crossing of the N = 126 shell closure is considered to be ubiquitous, having previously been observed (or inferred where measurements N = 125 – 127 have not been made) for Z = 81 – 84, 86 – 88. However, as discussed in the introduction to this chapter, the only even Z chain of charge radii measurements where N = 124, 126, 128 have previously been published is the lead (Z = 82) isotopes. The values of $R(m_{(2N)})$ and $R(m_{(1N)})$ calculated for mercury and lead are presented in Table 10.2, together with the kink predicted by covariant Density Functional Theory (DFT) calculations $R(m_{(2N_DFT)})$ from [13].

It is clear from the values in Table 10.2, that the prediction of the covariant DFT calculations overestimate the gradient of the kink defined as $R(m_{(2N)})$. The uncertainty of $R(m_{(2N)})$ for mercury is <3%, sufficient for comparisons with nuclear models which consider an agreement to within 10% to be reasonable [5]. As expected, there is a significant difference between the values of $R(m_{(1N)})$ and $R(m_{(2N)})$ due to odd-even staggering contributions.

The kinks in the mercury and lead isotope chains are plotted in Fig. 10.4, together with the predictions from Droplet model calculations using the parametrization from [14] and the results of covariant DFT calculations reported in [13]. Both sets of calculations are fixed to the experimental data points at N = 126 and the isotope chains are arbitrarily displaced to facilitate comparison. The measured and pre-

Table 10.2 The measured $R(m_{(2N)})$, the predicted $R(m_{(2NDFT)})$ from [13] and the measured $R(m_{(1N)})$ in the mercury isotope chain at the crossing the N = 126 shell closure. The measured values are from this work (Hg: N = 126, 127, 128), calculations of the $\delta\nu^{198,A}$ reported in [7] (Hg: N = 124, 125) and from [1] (Pb: N = 124 – 128)

Element	$R(m_{(2N)})$	$R(m_{(2NDFT)})$	$R(m_{(1N)})$
Hg	1.980 (52)	2.86	1.267 (60)
Pb	1.787 (11)	2.51	1.271 (8)

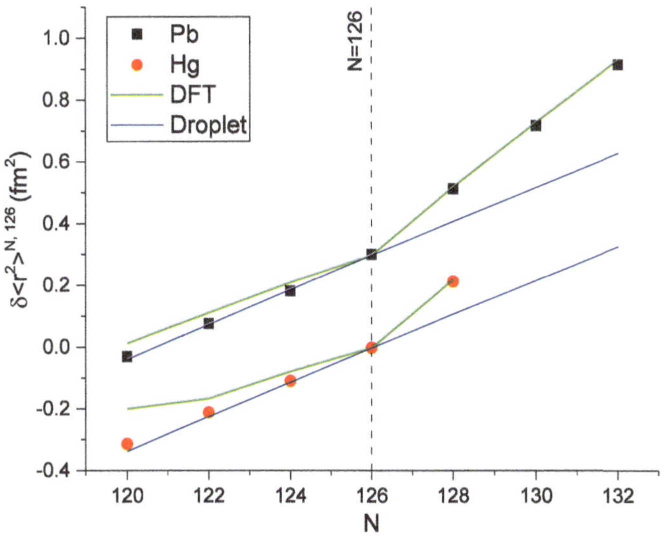

Fig. 10.4 Isotope shifts of lead and mercury isotopes across the N = 126 shell closure, with the predictions of covariant Density Functional Theory (DFT) calculations from [13] and Droplet model calculations using the parametrization of [14]. The experimental data points are taken from this work (Hg: N = 126, 128), calculations of the $\delta\nu^{198,A}$ reported in [7] (Hg: N = 120, 122, 124,) and from [1] (Pb: N = 120, 122, 124, 126, 128). The lead and mercury isotopes are offset from each other for clarity

Table 10.3 $m_1^{(2N)}$ and $m_2^{(2N)}$ gradients for mercury and lead isotopes presented in Fig. 10.4, together with the predictions of [13] ($m_1^{(2NDFT)}$, $m_2^{(2NDFT)}$). The gradients are defined in the text

Element	$m_1^{(2N)}$ (fm^2)	$m_1^{(2NDFT)}$ (fm^2)	$m_2^{(2N)}$ (fm^2)	$m_2^{(2NDFT)}$ (fm^2)
Hg	0.1080 (22)	0.0769	0.2140 (35)	0.2203
Pb	0.1179 (5)	0.0881	0.2107 (9)	0.2209

dicted gradients $m_1^{(2N)} = \delta\langle r^2\rangle^{N=124,126}$ and $m_2^{(2N)} = \delta\langle r^2\rangle^{N=126,128}$ are also given in Table 10.3.

Considering Fig. 10.4, the Droplet model is capable of reproducing the trend of isotope shifts in both the mercury and lead isotope chains in the immediate vicinity below the N = 126 shell closure. However, it fails to reproduce the kink. This is to be expected as the Droplet model is not capable of reproducing shell effects and is based on a set of empirically determined parameters, this is a clear demonstration of the limitations of this model. Considering Fig. 10.4 and Table 10.3, there is a discrepancy between the covariant DFT calculations and the trend of the lead and mercury isotopes below the N = 126 shell closure however, above the shell closure the covariant DFT calculations are well able to reproduce the trend in isotope shifts for both mercury and lead isotopes.

10.5 Conclusion

These measurements have enabled $R(m_{(2N)})$ to be determined for the mercury isotope chain. Used in combination with the previous measurements of lead, this will enable the benchmarking of the capability of nuclear models to reproduce shell effects in the heavier region of the nuclear chart. The results presented here were employed in this manner in [8].

References

1. Angeli I, Marinova KP (2013) Table of experimental nuclear ground state charge radii: an update. At Data Nucl Data Tables 99(1):69–95. https://doi.org/10.1016/j.adt.2011.12.006
2. Stone JR, Reinhard P-G (2007) The Skyrme interaction in finite nuclei and nuclear matter. Prog Part Nucl Phys 58(2):587–657. https://doi.org/10.1016/j.ppnp.2006.07.001
3. Sharma MM, Lalazissis GA, Ring P (1993) Anomaly in the charge radii of Pb isotopes. Phys Lett B 317(1–2):9–13. https://doi.org/10.1016/0370-2693(93)91561-Z
4. Kreim K, Bissell ML, Papuga J, Blaum K, De Rydt M, Garcia Ruiz RF, Goriely S, Heylen H, Kowalska M, Neugart R, Neyens G, Nörtershäuser W, Rajabali MM, Sánchez Alarcón R, Stroke HH, Yordanov DT (2014) Nuclear charge radii of potassium isotopes beyond N=28. Phys Lett B 731:97–102. https://doi.org/10.1016/j.physletb.2014.02.012
5. Goddard PM, Stevenson PD, Rios A (2013) Charge radius isotope shift across the N = 126 shell gap. Phys Rev Lett 110(3):032503. https://doi.org/10.1103/PhysRevLett.110.032503
6. Barzakh AE, Fedorov DV, Ivanov VS, Molkanov PL, Moroz FV, Orlov SY, Panteleev VN, Seliverstov MD, Volkov YM (2018) Shell effect in the mean square charge radii and magnetic moments of bismuth isotopes near N=126. Phys Rev C 97(1):014322. https://doi.org/10.1103/PhysRevC.97.014322
7. Ulm G, Bhattacherjee SK, Dabkiewicz P, Huber G, Kluge H-J, Kühl T, Lochmann H, Otten EW, Wendt K, Ahmad SA, Klempt W, Neugart R (1986) Isotope shift of 182Hg and an update of nuclear moments and charge radii in the isotope range 181Hg-206Hg. Z Phys Hadron Nucl 325(3):247–259. https://doi.org/10.1007/BF01294605
8. Goodacre TD, Afanasjev AV, Barzakh AE, Marsh BA, Sels S, Ring P, Nakada H, Andreyev AN, Van Duppen P, Althubiti NA, Andel B, Atanasov D, Billowes J, Blaum K, Cocolios TE, Cubiss JG, Farooq-Smith GJ, Fedorov DV, Fedosseev VN, Flanagan KT, Gaffney LP, Ghys L, Huyse M, Kreim S, Lunney D, Lynch KM, Manea V, Palenzuela YM, Molkanov PL, Rosenbusch M, Rossel RE, Rothe S, Schweikhard L, Seliverstov MD, Spagnoletti P, Van Beveren C, Veinhard M, Verstraelen E, Welker A, Wendt K, Wienholtz F, Wolf RN, Zadvornaya A, Zuber K (2021) Laser spectroscopy of neutron-rich 207,208Hg isotopes: illuminating the kink and odd-even staggering in charge radii across the N = 126 shell closure. Phys Rev Lett 126(3):032502. https://doi.org/10.1103/PhysRevLett.126.032502
9. Fink DA, Cocolios TE, Andreyev AN, Antalic S, Barzakh AE, Bastin B, Fedorov DV, Fedosseev VN, Flanagan KT, Ghys L, Gottberg A, Huyse M, Imai N, Kron T, Lecesne N, Lynch KM, Marsh BA, Pauwels D, Rapisarda E, Richter SD, Rossel RE, Rothe S, Seliverstov MD, Sjödin AM, Van Beveren C, Van Duppen P, Wendt KDA (2015) In-source laser spectroscopy with the laser ion source and trap: first direct study of the ground-state properties of 217, 219Po. Phys Rev X 5(1):011018. https://doi.org/10.1103/PhysRevX.5.011018
10. Dabkiewicz P (1980) Laserspektroskopische bestimmung der formisomerie in 185Hg und der isotopieverschiebung von 206Hg. PhD thesis, The Johannes Gutenberg University of Mainz
11. Fricke G, Heilig K (2005) 80-Hg mercury. In: Nuclear charge radii. Springer, Berlin, pp 1–9. https://doi.org/10.1007/10856314_82

12. Dabkiewicz P, Duke C, Fischer H, Kuehl T, Kluge HJ (1978) Nuclear shape staggering in very neutron deficient Hg isotopes detected by laser spectroscopy. J Phys Soc Jpn 44(suppl):503–508
13. Agbemava SE, Afanasjev AV, Ray D, Ring P (2014) Global performance of covariant energy density functionals: ground state observables of even-even nuclei and the estimate of theoretical uncertainties. Phys Rev C 89(5):054320. https://doi.org/10.1103/PhysRevC.89.054320
14. Berdichevsky D, Tondeur F (1985) Nuclear core densities, isotope shifts, and the parametrization of the droplet model. Z Phys Hadron Nucl 322(1):141–147. https://doi.org/10.1007/BF01412027

Chapter 11
Conclusion

This thesis presents both a series of developments for the ISOLDE RILIS and the results following their application for a study of the ground state properties of exotic mercury nuclei.

Chapter 6 presents the identification and characterization of a new laser-atom interaction region, coupling the RILIS with the VADIS, ISOLDE's variant of the FEBIAD-type arc discharge ion source. This has resulted in a truly versatile ion source the VADLIS (Versatile Arc Discharge and Laser Ion Source). This development has enabled the application of the RILIS with molten metal targets at ISOLDE, increased the flexibility of ISOLDE operation and opened a new direction for ion source development. Chapter 7 presents the development of new RILIS ionization schemes for chromium, germanium, mercury, radium and tellurium. In the course of the radium scheme development, autoionizing states were identified for the first time, in addition to the observation of optical pumping in the hot cavity environment.

The VADLIS development and the new RILIS ionization scheme for mercury were instrumental in the success of the in-source RIS of exotic mercury isotopes presented in Chaps. 8–10. The increase in the ionization efficiency of a factor of 60 (compared to the previous RILIS ionization scheme) together with the sensitivity of the technique, enabled HFS measurements on the mercury isotope chain to be extended four isotopes further towards the proton drip line: down to ^{177}Hg and two isotopes further into the neutron rich region of the nuclear chart up to ^{208}Hg.

Analysis of the extracted mean square charge radii definitively determined ^{180}Hg to be the end point of the famous odd-even staggering in the light mercury isotopes. The reduction in deformation offers possible evidence for the robustness of a neutron shell closure around $N = 82$, well beyond the proton drip line.

The extension of the isotope shift measurements at the neutron rich end of the mercury isotope chain, corresponded to the crossing of the $N = 126$ neutron shell closure. This enabled the determination of the characteristic kink in the gradient of the isotope shifts. This is the lowest Z isotope chain for which the mean square charge radii of isotopes with $N = 127$ and $N = 128$ has been studied and only the third overall

T. Day Goodacre, *Applied Laser Spectroscopy for Nuclear Physics*, Springer Theses,
https://doi.org/10.1007/978-3-030-73889-1_11

after lead (Z = 82) and bismuth (Z = 83). Measurement of the N = 124, 126 and 128 isotopes is essential for the most accurate determination of the kink, as it enables the change in the gradient of the isotope shifts to be determined across the shell closure without the influence of odd-even staggering effects. In this region of the nuclear chart, it is often only the even-even isotopes that are accessible to nuclear theory calculations. Thus these results are particularly applicable for the benchmarking of nuclear theory calculations as they attempt to reproduce shell effects in the trend of mean square charge radii.

About the Author

The author completed a BSc. in Mathematics at the University of Leeds, and then an MSc. in Photon Science at the University of Manchester. Following this, he undertook a three-year Marie Curie Early Stage Research Fellowship based at CERN (the European Organization for Nuclear Research) while working on a PhD. in Nuclear Physics with the University of Manchester. After the completion of the PhD., he joined TRIUMF (Canada's Particle Accelerator Centre) as a Postdoctoral Research Fellow, and subsequently became a staff scientist focused on the development of TRIUMF's Advanced Rare Isotope Laboratory. Thomas is currently the Group Leader for Target and Ion Source R&D at TRIUMF, an up to date list of his publications can be found at https://orcid.org/0000-0003-1499-0888.

T. Day Goodacre, *Applied Laser Spectroscopy for Nuclear Physics*, Springer Theses,
https://doi.org/10.1007/978-3-030-73889-1

The manufacturer's authorised representative in the EU is Springer Nature Customer Service Centre GmbH, Europaplatz 3, 69115 Heidelberg, Germany. If you have any concerns regarding our products, please contact ProductSafety@springernature.com

Printed and bound by CPI Group (UK) Ltd, Croydon, CR0 4YY
28/09/2026
02224089-0017